VIRTUALLY INTEGRATED HEALTH SYSTEMS

VIRTUALLY INTEGRATED HEALTH SYSTEMS

A Guide to Assessing Organizational Readiness and Strategic Partners

Richard J. Coffey
Kate M. Fenner
Sheryl L. Stogis

Jossey-Bass Publishers
San Francisco

Substantial discounts on bulk quantities of Jossey-Bass books are available to corporations, professional associations, and other organizations. For details and discount information, contact the special sales department at Jossey-Bass Inc., Publishers (415) 433–1740; Fax (800) 605–2665.

For sales outside the United States, please contact your local Simon & Schuster International Office.

Jossey-Bass Web address: http://www.josseybass.com

TCF Manufactured in the United States of America on Lyons Falls Turin Book. This paper is acid-free and 100 percent totally chlorine-free.

Library of Congress Cataloging-in-Publication Data

Coffey, Richard James.
    Virtually integrated health systems : a guide to assessing
organizational readiness and strategic partners / Richard J. Coffey,
Kate M. Fenner, Sheryl L. Stogis. — 1st ed.
        p.   cm.
    Includes bibliographical references and index.
    ISBN 0-7879-1078-3 (alk. paper)
    1. Integrated delivery of health care.   2. Health services
administration.   I. Fenner, Kathleen M.   II. Stogis, Sheryl L.
III. Title.
    [DNLM: 1. Delivery of Health Care. Integrated—organization &
administration.   W 84.1 C674v 1997]
    RA971.C66   1997
    362.1′068—dc21
    DNLM/DLC
    for Library of Congress   97-39788

FIRST EDITION
HB Printing   10 9 8 7 6 5 4 3 2 1

CONTENTS

PREFACE

The healthcare industry is in the midst of a tectonic shift. Major employers and governmental agencies are taking actions to reduce the costs, or at least reduce the inflation of costs, of healthcare services, and insurance holders are demanding greater coverage of services. In response to these pressures, managed care organizations have formed as mechanisms to provide a single contract with employers; limit expenditures through the selective use of lower-cost physicians, hospitals, and other healthcare organizations; and reduce the use of healthcare services. Because of the decreasing demand for inpatient hospital services, the need to control costs, the need to be included in the provider panels of managed care organizations, and increasing competition among healthcare providers, healthcare organizations are joining together through a number of different structural arrangements, among them alliances, mergers, acquisitions, and joint ventures. These arrangements may be not-for-profit, for-profit, or hybrids. But whether the organization is not-for-profit or for-profit, the need to provide high-quality, cost-effective care and to produce a profit remains critical. Both for-profit and not-for-profit organizations reinvest much of the profit to renew and change their facilities, equipment, and services; for-profit organizations, however, direct part of the profits to the stockholders, whereas not-for-profit organizations use that money within the communities they serve.

To gain more control over and better coordinate the wide range of resources, health systems or networks are becoming larger and more integrated. A number of different terms are used to describe these organizations. We use the terms *virtually integrated health systems* or *integrated health systems*. We have consciously referred to the combined terms *healthcare, health, and social services* instead of *healthcare* alone because the emerging systems will place more emphasis beyond traditional healthcare services on services that promote health. We have also used the term *virtually integrated* to indicate that collaborating organizations can function effectively as integrated entities.

Purpose of the Book

Virtually Integrated Health Systems provides insights, approaches, and tools based on our many years of experiences with a wide variety of organizations. Our goal is to provide direction, approaches, and tools to help you assess your organization, existing or potential partners, and competitors in relation to integrated health systems. Where does your organization fit in a large, integrated health system? What strengths do you offer an integrated system, and what are your current weaknesses? How can you best present your organization to potential partners? To help you answer these and other questions, we explore approaches and tools to help you make assessments related to the scope of services, external expectations and criteria, and the internal organizational climates. You can use these same tools as a basis for discussions regarding the roles of different systems partners and to assess your collective strengths compared to those of your competitors.

Our specific aim is that everyone who reads this book will gain substantial value from the ideas, approaches, and tools we examine. By recognizing and addressing issues early, you may avoid huge conflicts and potential merger or acquisition failures that may cost millions of dollars. At the other extreme, even if you only use a few of the ideas from this book or it only contributes to a better understanding of the changes, it should provide a return on investment many times its cost.

Audience

Leaders and managers of healthcare, health, and social services organizations are the primary audience for *Virtually Integrated Health Systems*. The book is expressly written for those considering, already developing, or currently operating an integrated health system. The audience is specifically intended to include people in nontraditional healthcare, health, and social services organizations; managed care companies and other payers; and employers, in addition to those considered part of the traditional medical care system. This audience includes anyone work- ing to improve or sustain health. The roles of these people will be increasingly im- portant as we shift more attention toward health promotion and maintenance.

The secondary audiences are students in a wide range of healthcare man- agement, health policy, medicine, nursing, social services, business, engineering, and other professions relating to healthcare, health, and social services. This book provides a practical understanding of the issues and important considerations to assess or form a virtually integrated health system (VIHS). The future of our so- ciety depends on future leaders' understanding how to form and operate virtually integrated health systems to improve the health of the communities they serve.

Leaders and managers in other organizations—governmental agencies, pro- fessional organizations, review and accreditation organizations, business and in- dustrial organizations, universities, law and accounting firms, and consulting companies—may also find the approaches and tools in this book helpful. They will gain an understanding of approaches and tools for assessing healthcare or- ganizations and integrated health systems and will then be able to adapt these ap- proaches and tools to merger situations in other industries.

Overview of the Contents

In Chapter One, we introduce the virtually integrated health system, which we believe is the health and social system of the future. In particular, we emphasize the need for a much broader view of healthcare, health, and social services than that of the traditional medical care model, which focuses on hospitals, outpatient clinics, and physicians. We explain the incentives under different reimbursement systems, so you can understand the conflicting pressures on different organizations and people. We are in the midst of a shift from a paradigm of sickness to a para- digm of wellness.

Chapter Two introduces the analysis approach that will be used throughout the book and describes the different scopes and uses of the analysis models. Spe- cific applications of the analysis tool are described in Chapters Four, Six, and Eight.

Chapter Three presents a taxonomy, or classification system, of healthcare, health, and social services. The taxonomy has six major dimensions: social and environmental conditions, health-related human conditions, foci, settings, core/key processes, and resources.

The analysis model described in Chapter Four provides a specific tool to help organizations assess themselves, potential partners, and competitors in relation to the scope of services. This analysis builds on the taxonomy presented in Chap- ter Three.

Chapter Five describes how external expectations and evaluation criteria can be used as an important part of the assessment. The chapter calls for consideration of a virtual set of expectations for healthcare, health, and social services organizations and demonstrates the feasibility of such a move. All of us have experienced or heard stories about external reviewers and review processes that do not add value. This chapter moves beyond the external review itself, addressing the commonality and value of external expectations and criteria to assess and improve an organization or an integrated health system.

An assessment tool related to external expectations is described and illustrated in Chapter Six. This assessment is based on the criteria common to most external assessment and accreditation organizations and uses the same format as described in Chapter Two and Four.

Chapter Seven addresses the vitally important internal organizational climate by describing several cultural and human dimension issues. Failures to integrate healthcare and health-related organizations most often stem from cultural, leadership, and human relations issues.

Chapter Eight illustrates an assessment tool related to the internal organizational climate, using the cultural and human dimensions described in Chapter Seven. Again, the same format is used for the assessment tool.

Chapter Nine describes some of the most common problems, or speed bumps, encountered on the road to integration, along with suggestions of how to avoid them. Attending to these issues early will greatly increase the likelihood of successful integration.

Since the three major types of analyses are interrelated, Chapter Ten describes an approach to consolidate these analyses through the use of a prioritization matrix. It explains how to create a balanced scorecard, with weights appropriate to a particular organization's situation. However, caution should be used when assigning weights, because it can cause inadequate attention to important criteria.

Chapter Eleven describes how all of these concepts and tools can be incorporated into organizational planning. It also makes suggestions on how to use the information from the assessments.

Comparisons ("crosswalks") of external review bodies' expectations and a comprehensive assessment tool are presented as appendixes.

Acknowledgments

Many people contributed to the creation of this book, and we wish to thank them all. A few we will be able to acknowledge personally, but it is futile to try to acknowledge all of our colleagues and clients who contributed to our knowledge and experience over the years that allowed us to create this book. Melissa Basford, at Compass Group, Inc., served as our executive editor and did a tremendous job of consolidating three different writing styles from frantically busy authors in three different states. Peter Fenner at Compass Group regularly asked questions and offered ideas that improved our thinking and this book. James Mason, president of Beech Acres, was helpful in testing our approaches related to child welfare and social services organizations. Our colleagues and clients provided the examples that we have used throughout the book, as well as reviews of different portions of the manuscript. During preparation of the manuscript, we were assisted in our research and editing by Misty Fairbanks.

We thank our friends and colleagues who reviewed drafts of the manuscript and helped us make the book more useful: Diane Bechel, Ford Motor Company; Julie Hanser, Mercy Regional Health System; Ellen Gaucher, Wellmark, Inc.; David Gurule, Group Health of Puget Sound; Pat Lyons, University of Michigan Health System; James Mason, Beech Acres; Sister Joanne Schuster, Franciscan Sisters of the Poor Health System; Tom Wilburn, TriHealth, Inc.; and Gary Young, Boston University. They provided valuable critiques and ideas to improve the book.

We thank Andy Pasternack, senior editor, and the editorial staff at Jossey-Bass, who provided tremendous support to us.

Finally, we thank our families, who were understanding of our passion to write this book in the midst of our hectic lives. We appreciate their support. Peter Fenner in particular served as a vitally important sounding board and encouraged us to push ahead when we were running out of energy. Our spouses, children, nieces, nephews, many sisters and brothers, other relatives, Pew6 colleagues, and friends demonstrated patience while we were working on this book, although they missed our attention. We dedicate this book to the key supporters in our families: Todd and Tonya Coffey, Peter Fenner and Tracie Abbott, and Ben and Pat Stogis.

September 1997

Richard J. Coffey
Ann Arbor, Michigan

Kate M. Fenner
Cincinnati, Ohio

Sheryl L. Stogis
Elk Grove Village, Illinois

THE AUTHORS

Richard J. Coffey, Ph.D., is director of program and operations analysis at the University of Michigan Health System, Ann Arbor, Michigan; an adjunct associate professor in industrial and operations engineering at the University of Michigan; and a consultant with Compass Group, Inc., Cincinnati, Ohio. He received a B.S.E. degree (1967) from the University of Michigan in industrial engineering and an M.S. degree (1971) from the University of Arizona in systems engineering. He holds an M.S.E. degree (1972) and a Ph.D. degree (1975), both from the University of Michigan, in industrial and operations engineering. Current activities include a multiyear effort to improve cost effectiveness and quality at the University of Michigan Health System. He is an investigator on a National Science Foundation Transformations to Quality Organizations grant and a Pfizer Health Research Foundation grant in Japan. He has served in staff, consulting, and leadership roles in many university and community healthcare organizations, consulting, government, insurance, and private organizations within the healthcare industry since 1963, both in the United States and abroad. He has authored or coauthored over forty-five publications. *Total Quality in Healthcare: From Theory to Practice* received the 1993 book-of-the-year award from the *American Journal of Nursing* and *Transforming Healthcare Organizations: How to Achieve and Sustain Organizational Excellence* received the 1992 James A. Hamilton book-of-the-year award from the American College of Healthcare Executives.

Kate M. Fenner, Ph.D., has twenty years of experience as a project coordinator and lead consultant in the healthcare environment. She is the president of Compass Group, Inc., and is a faculty member for the Joint Commission on Accreditation of Healthcare Organizations. She received her B.S. degree (1970) in nursing from Northern Illinois University and her M.S. degree (1973) in community health nursing. She holds a Ph.D. degree (1979) in ethics and law in healthcare from the Union Graduate School, Cincinnati, Ohio. Her dissertation became a leading college textbook on the legal and ethical aspects of healthcare. Fenner has served in a variety of academic leadership roles, as professor, college dean, and university vice president. She has won acclaim for her presentations on the changing nature of healthcare targeted to healthcare and physician leaders. In addition to numerous articles, she has published two books: *The Manual of Nursing Recruitment and Retention,* written with Peter Fenner, has been used as the foundation for many subsequent texts, and *Ethics and Law in Nursing: Professional Perspectives* is used in higher education.

Sheryl L. Stogis, Dr.P.H., is the CEO of the Michigan Peer Review Organization. She has over twenty years of healthcare experience in education, quality management, marketing, operations redesign, consulting, project management, nursing, and sales. In 1994, she became an award-winning Pew fellow. She holds a B.A. degree in English and writing from Illinois Wesleyan University, a B.S. in nursing from Northwestern University, a master's degree in public service from De Paul University, and a doctorate in health policy from the University of Michigan. She was a team leader for the Regional Organ Bank of Illinois and a consultant and faculty member for Compass Group, Inc. Stogis spent several years with the Joint Commission on Accreditation of Healthcare Organizations; she worked on the Orion Project, which redesigned the entire accreditation and operations process, and was responsible for the management, training, and support of hospital and ambulatory surveyors. She also worked on the measurement system and served as faculty for various educational programs on quality and leadership.

VIRTUALLY INTEGRATED HEALTH SYSTEMS

CHAPTER ONE

EMERGING HEALTH AND SOCIAL SYSTEMS

A virtually integrated health system (VIHS) interweaves community resources to address the continuum of health requirements for a defined population using the least intensive, highest-value strategy. This expanded definition of health with an emphasis on prevention means that new players will be brought to the table. Social services and child welfare advocates, houses of worship, and educational resources will all be involved in a comprehensive community-based approach to health promotion and maintenance.

The term *virtually integrated health system* covers a variety of strategies available for the design and delivery of health services to a defined population. Virtual integration indicates that there is no single, organized health system but rather a network of health and social service organizations, working together with a community or population, that adapts to achieve a desired effect. The critical understanding required for effective design and delivery of healthcare is to begin with an understanding of the health needs of the population or community to be served. Needs assessment thus drives system design.

In fact, a comprehensive delivery system addressing the full range of assessed needs of a defined population in the variety of settings, services, and foci required is most likely not in existence. In addition, developing such a system as a wholly owned corporate entity would be very costly, duplicative of existing services, and unwieldy to manage. The crafting of a VIHS will require stretching the usual organizational definitions and boundaries to embrace the multitude of potential participant providers of service to the population served. Virtually integrated ways of providing services will need to be explored, among them the formation of alliances, federations, and cooperative ventures between heretofore unfamiliar providers or provider segments. The prevalent wholly owned and tightly controlled contemporary model will be stretched and changed through new relationships and arrangements and will require new skills and abilities to cooperate and influence without control and ownership.

In healthcare, the 1990s may come to be remembered as the decade of consolidation and integration (Stefl, 1996, p.1). In the rush to establish integrated health systems, many organizations have made costly, and in some cases disastrous, errors in planning and implementation. Millions of dollars have been invested in mergers that fall apart before they are completed due to irreconcilable differences. Readers of *Modern Healthcare* may recall the failed merger of Community and St. Vincent's hospitals in Indianapolis. After the investment of millions of dollars toward integration, the deal failed when the respective boards could not agree on selection of executive leadership.

Every person and organization relating to the healthcare industry is facing a rapidly changing environment. The skyrocketing increases in medical care costs have laid the groundwork for the growth of managed care. Between 1978 and 1994, the enrollment in health maintenance organizations (HMOs) rose more than 450 percent. Between 1993 and 1994 alone, HMO penetration grew 12 percent, with Medicare and Medicaid accounting for 24 percent and 42 percent, respectively, of the increase (National Committee for Quality Health Care, 1997). Payers, including the federal and state governments and the rapidly growing managed care organizations, are demanding greater cost-effectiveness.

There has been an astonishing 9 percent decrease in the number of hospitals since the 1970s. The number of community hospitals fell by more than 11 percent between 1980 and 1994, and almost 30 percent of those were state and local government community hospitals (American Hospital Association, 1995). In the sixteen-year span between 1978 and 1994, the total number of community hospital beds fell more than 20 percent (National Committee for Quality Health Care, 1997). But even with these decreases, there remains a surplus of hospital beds. Providers are responding by providing more services in ambulatory, home, and other noninpatient settings. In addition, new technologies have allowed shorter lengths of stay and added more services provided in the ambulatory setting. "The number of inpatient admissions per 100,000 population and average length of stays have fallen by about 24 percent and 12 percent, respectively. By contrast, less resource-intensive types of utilization have soared. In 1994, the numbers of ambulatory surgeries, home health visits, and hospice days per 100,000 population were each at least five times their 1978 level" (p. 10).

Another response from healthcare providers has been to reduce the skill mix and the number of staff. In the 1980s, one in every six new jobs was created in the healthcare sector. In 1993, medical care employment accounted for 8.4 percent of all jobs. With the increase in health spending slowing, some experts predict a loss of 600,000 to 1 million new jobs (Auerbach, 1996). At the same time, patients and other customers, sensitized to quality service provided by world-class organizations in other industries, and at the same time afraid that healthcare as they know it will be taken out of their control, are demanding improved service.

All of these trends have had a direct impact on the healthcare system. Beyond these traditional indicators are other trends that ultimately will have an impact on the emerging health system:

- Every forty-seven seconds, a child is abused or neglected.
- Every thirty-six minutes, a child is killed or injured by a gun.
- Healthcare coverage remains nonexistent for more than 38 million Americans. Of those 38 million, 9.8 million are children.
- More than 1 million females under the age of twenty, or one in ten, become pregnant each year. Between 30 and 40 percent of teenagers who became pregnant will conceive again within two years.
- Infants and children with medical problems and physical or mental limitations are the fastest-growing segment in foster care. Many of these children's health and mental health needs are not being met.

Communities often provide a piecemeal set of social services interventions, and the current delivery system is fraught with both gaps in necessary services and duplicative, overlapping services that are expensive to deliver. Health and welfare system reform must create incentives for the efficient and effective delivery of services to children and families.

Using a medical model, it is easy to see that a child with cancer will either be cured or die, but that cancer is not spread to others by the child. An abused, neglected child, however, can be viewed like a carrier of a deadly, infectious disease. That child will carry the disease of violence and despair into future generations, proliferating the devastation at a continuing, and growing, cost to all.

This is the age of integration and consolidation as healthcare organizations attempt to establish some control over the market, which is currently controlled by managed care organizations and others who control the patients. We now have a unique opportunity to bring to the table not just healthcare organizations but those other organizations that can also benefit from refocusing on prevention of healthcare costs through community-based services.

The question is how to address these concerns and create the most effective structure for providing health, healthcare, and social services. Should we patch the system now and wait for the next round of major changes, or should we engage in long-term strategic planning of a health and social services system that will pay us back in lower costs and higher quality of life for the community over the next generations? Our excuse thus far is that we in America want immediate results, so it is impossible to wait for the payoff of prevention. *The payoff for lack of prevention will make this era of capitation look like a picnic.* The instability of the current environment can promote collaboration between entities within a community, emphasizing a more horizontal plan for health and social system development. Discussion and partnerships between these and other organizations is a viable alternative to a trend in decreased government spending on social services concomitant with capitated healthcare dollars. Never before has there been a greater opportunity to focus and direct investments in prevention.

We begin by defining some of the issues, including financial incentives and reimbursement. In subsequent chapters, we propose an approach to help organizations craft a VIHS that draws on all the resources in the community. The aim is to cause people to think outside their normal frameworks and consider nontraditional alternatives. Hospital leaders, for example, must include in their own thinking nontraditional health services, child welfare services, social services, and other services that affect the health of a community's population. Similarly, social services agencies must view themselves as part of the VIHS. A broader consideration of healthcare, health, and social services permits the management of acute care services.

The evolution to a virtual system is not easy. The integration of administrative functions and disparate clinical functions faces many challenges, and most systems are only now realizing the scope and enormity of the task. Every system may choose to affiliate in different ways. There are many variants of system components, ownership, level of asset merger, administrative authority, physician integration, contractual relationships, and types of employment. Hence, there are no simple, commonly used definitions. Following are some examples of types of affiliations and systems:

- *Alliances.* This is probably the earliest and most common form of integration into a loose system. It is a group of healthcare organizations that have formed together for some reason or purpose. Many times the initial purpose was to establish purchasing power. Now several of these alliances have branched out to include clinical information development and the sharing of systems. Alliances have no central ownership or administrative authority. Examples of this type of system are Volunteer Hospitals of America and Premier.

- *Multiple hospital systems.* Two or more hospitals join together and commonly are networked within a defined geographic region. Typically these are name-only types of agreements; they cover a range of ownership arrangements and commonly include referral arrangements.
- *Physician organization systems.* This arrangement aligns physician groups and organizations in an economic or strategic merger.
- *Networks.* This arrangement can be similar to alliances, but with the tendency to include multiple types of healthcare organizations. These structures can offer a broader scope of services. The affiliation can be deeper than the name-only hospital systems and may include mergers and acquisitions.
- *Organized delivery system.* One definition of this system is "a network of organizations that provides or arranges to provide a coordinated continuum of services to a defined population and is willing to be held clinically and fiscally accountable for the outcomes and the health status of the population served" (Shortell and others, 1996, p. 7).
- *Virtually integrated health systems.* A group of healthcare, health, and social services organizations that provides or arranges to provide a broad continuum of services to improve the health of community residents and is accountable clinically and financially for the outcomes and health status of the community population. The primary difference from an organized delivery system is that a VIHS provides a broader scope of services and activities oriented toward the promotion, maintenance, and enhancement of health, in addition to the traditional services to address illness and injuries.

Although the terminology varies, the virtually integrated health system or network is a mechanism whereby organizations provide one-stop shopping for patients and complete services and geographic coverage for exclusive provider relationships with employers. The ownership and structure of these organizations vary widely and are of less importance to customers than achieving seamless delivery of a broad range of insurance and health services. Through ownership or agreements, VIHSs normally include the following components:

- An insurance company that functions as a managed care organization
- A range of health insurance products
- Acute care hospitals, including one or more medical centers
- A broad range of ambulatory care services
- Physicians employed by or closely linked to the system
- Inpatient alternatives to acute care hospitals, including subacute care facilities or skilled nursing facilities
- Home-, work-, or school-based services

Healthcare, Health, and Social Systems

The traditional healthcare system, with a focus on medical care to treat selected illnesses and injuries, is too narrow to address health and quality of life fully. The major emphasis of the system has been on hospitals, which provide the most critical and costly care, with less focus on outpatient services and physicians' offices and very little attention on other services related to health, such as school health programs, child and family services, drinking and driving, and violence. The traditional healthcare system diagnoses and treats survivors of auto accidents resulting from drunk driving, but does little to work with other organizations in the community to prevent the accidents in the first place or minimize the injuries resulting from the accident.

The continued existence of all organizations that provide healthcare, health, and social services will depend on understanding the environment, the competition, and the alternatives to provide value in the future. The analytical approaches and tools presented in this book can help structure your understanding of your organization, your current and potential partners, and your competitors.

A New Conceptual Model: From Sickness to Wellness

All of these changes are causing increasing numbers of healthcare professionals, employers, and community residents to focus more on maintaining health than treating sickness. Our past views, approaches, and even goals are changing. We are switching from an acute care focus on keeping people physically alive to a focus on keeping people healthy and providing required care in the least costly setting possible. Quality of life is being considered in addition to the sheer duration of physical life. A paradigm shift is creating a whole new way of thinking within the healthcare industry.

Health has been defined in large part as the absence of physical illness, injury, or disability, and the goal has been to keep people physically alive as long as possible. But health is more than that; a broader definition is the collective state of physical, mental, social, and spiritual well-being. Healthcare, or more accurately medical care, has in large part been reactive in nature. If people had some physical, or to a lesser degree mental, problem, they went to a physician or a hospital. Although this type of care, reactive to physical needs, is very effective for people once they reach that state, this reactive approach is inadequate to prevent the underlying causes of the illness or injuries. Consider the following examples:

- Physicians and hospitals treat child abuse cases but need more proactive partnerships with others in the community to prevent child abuse.
- Physicians and hospitals treat injuries from auto accidents, but should promote more active community involvement to prevent drinking and driving.
- Physicians and hospitals treat gunshot and knife wounds but do little to address accident or violence prevention. Once a person has a spinal cord injury from an accident, he or she faces a life of disability and large medical costs.
- Physicians and hospitals treat lung cancer, but do little to reduce smoking.
- Physicians and hospitals treat heart disease, but need to team with others in the community, including food stores, to reduce risk factors associated with heart disease within the population.

These and multitudes of other similar situations are part of our mental model related to medical care or sickness. There is sort of a chicken-and-egg situation among what people and organizations choose to insure for, the incentives thus created by payers affect services provided, and the services, in turn, affect the health of the population.

Health, as we have defined it here, is much broader than the traditional medical care system and requires a different way of thinking. The boundaries and rules should change to address health, in addition to illnesses and injuries. Several contrasts between the focus of sickness and the focus of wellness are illustrated in Figure 1.1. As we begin to focus more on promoting health and preventing illness, rather than reactively treating illnesses and injuries, we face several shifts of thinking. John Griffith advocates for integrated healthcare organizations (IHCOs) to have a community rather than a membership orientation, and this will be a distinguishing characteristic and a source of market appeal. The transition to IHCOs will be slow; to prosper, the IHCO will have to accommodate both price-oriented markets and traditional ones (Griffith, Warden, Dowling, and Pelham, 1996, p. 4). Some question whether this community health focus, which we also advocate, will have sufficient market appeal to be financially viable.

Another major change will be the way we look at rights and responsibilities. In the United States, individual freedoms and rights are currently held above the subsequent public responsibilities and costs. Motorcycle helmets and handguns are easy examples. Individuals tenaciously demand the freedom and right to ride motorcycles without helmets and to own and carry handguns. Yet healthcare providers and the public are expected to be responsible for the care and costs related to subsequent injuries and deaths. If we as a society were to place greater emphasis on the health and rights of the public than those of individuals, then selected individual freedoms and rights may be curtailed, or individuals may be

FIGURE 1.1.   CONCEPTUAL MODEL: SICKNESS TO WELLNESS.

Description	Sickness Orientation	Wellness Orientation
Definition	• Sickness defined primarily in physical terms and in terms of absence of sickness • Continuation of physical life is goal	• Health includes all domains of life: physical, mental, social, and spiritual • Duration of quality life is goal
Basis	• Based on people who come for service	• Based on needs of population in community
Rights and responsibilities	• Individual freedoms and rights to risky behaviors, and professional and public responsibility to repair and pay for damage • Insurance premiums based on group	• Public rights to safe environment and individual responsibility for own behaviors • Tighter controls on public health issues (environment, smoking, etc.) • Insurance premiums based in part on personal behaviorally generated risk factors (excludes genetic and nonbehavioral factors)
Driver	• Driven by healthcare professionals • Healthcare providers as decision makers	• Driven by patients, covered families, and community residents • Healthcare providers are consultants; patients and families are decision makers
Reimbursement	• Reimbursement favors use of most expensive resources	• Reimbursement favors maintaining health and using most cost-effective services
Care approach	• Unquestioned "more is better"	• Collaboratively developed practice guidelines, protocols, and critical paths
Settings	• Settings are hospital and ambulatory medical care facilities (centralized)	• Many services provided in home, work, school, and community settings (decentralized)
Outcomes	• Outcomes measured in terms of short-term survival and complications	• Outcomes measured in terms of long-term functionality
Providers	• Emphasis on licensed medical care providers (doctors, nurses, therapists, etc.)	• Extends providers to include many community-based health-related people (chiropractors, coaches, clergy)
Information	• Providers own patient and personal information • Information retained separately by each provider	• Patients/clients own information, with limited exceptions • Availability of information across providers and settings for continuity of care

FIGURE 1.1.   (continued)

Description	Sickness Orientation	Wellness Orientation
	• Data systems based on billing information	• Data systems based on health and care requirements
Education	• Medical and health education focused on healthcare providers	• Broad education of population to manage health and basic care for self, family, and friends • Example is all high school students trained in basic first aid

held responsible for their known risky behaviors. One example would be charging higher health insurance premiums for smokers and alcoholics. Should this change occur, the public should decide on a limited set of such behaviors to prevent insurance companies from creating a complex set of punitive insurance premiums to their advantage. Also, health risks not controllable by a person (for example, congenital conditions, diabetes, and kidney failure) should not be allowed as a basis for insurance companies to discriminate against those people, because the people cannot control those health risks. Hospitals in particular must rethink their focus and role. One hospital that did this is Doctors Hospital in Toronto, Ontario. Although it had approved funding to build a new three-hundred-bed hospital, the leadership decided to downsize the inpatient facility replacement radically and focus instead on ambulatory services for its multiethnic community. Tertiary inpatients are referred to several close major Toronto medical centers. With the help of community advisory groups, several community-based clinics and services were established.

We are now in a transitional period, moving from the paradigm of sickness to the future paradigm of wellness. Selected transitions are illustrated in Figure 1.2. Note that silo-based organizations have rigid divisions and departments characterized by lack of communication, inattention, and an inward rather than customer-oriented focus.

Financing and Incentives

The scope of services offered, their utilization, and the behaviors of providers are closely linked to reimbursement systems.

FIGURE 1.2.   TRANSITIONS RELATED TO HEALTH AND SOCIAL SYSTEMS.

	Time Period		
Description	**Past**	**Present**	**Future**
Paradigm	Sickness Personal rights	Transition	Wellness Personal responsibility Quality of life
Driver	Professional disciplines	Payers	Employers Public
Service	Institutionalized	Mixed	Distributive health

Leadership	Top-down
	   professionally driven
			Dispersed client
			   driven

Behavior	Silos	Leaky silos	Integrated quilt
Information	Provider paper records Discrete event data		Shared, computer- ized, continuous data
Focus	Quality	Cost	Value and access

Types of Reimbursement Systems

There are essentially four different approaches to reimbursing for services related to healthcare, health, and social services.

Fee-for-Service Reimbursement. Under this approach, providers are reimbursed based on a fee schedule for the services they provide. The fee schedule may be established by the provider, the payer, or jointly. The point is that providing more services results in more reimbursement. Although payers negotiate discounts to the fee schedule, and sometimes large ones, the providers still receive more total reimbursement if they provide more services. This has been the traditional reimbursement mechanism for doctors and hospitals. Following are examples of fee-for-service reimbursement:

- Visiting nurses are reimbursed for each home visit.
- Speech therapists are reimbursed for time spent with patients.

- Hospitals are reimbursed for each day of hospital stay, each test performed, and every supply used. Internal medicine, pediatric, and some other physicians are reimbursed for each day of care while a patient is in a hospital.
- Physical therapists are reimbursed for each therapy session they provide.
- Child welfare or family services organizations are paid for hours spent with patients.

Later, we explain the incentives under this and the other reimbursement systems.

Fixed-Price-per-Episode Reimbursement. This approach provides a fixed reimbursement for a specific episode of care. The episode is defined differently by different payers and may not include what most would consider the entire episode of care for a given illness or injury. Again, the fee schedule may be established by the provider, the payer, or negotiated. Fixed-price-per-episode reimbursement is similar to purchasing a product in a store for a fixed price, independent of the amount, mix, or cost of resources required to produce that product. The following examples illustrate the fixed-price-per-episode approach:

- Some hospitals are paid a fixed price for patients discharged with certain diagnosis-related groups (DRGs). The episode is normally defined as the inpatient stay only and may be much shorter than the episode of illness or injury.
- Some surgeons are paid a fixed price for a surgical procedure. The episode is normally defined to include a specified amount of pre- and postsurgical care, including the time a patient spends as a hospital inpatient, and a limited number of follow-up outpatient visits.
- Some child welfare organizations receive a case-based rate for services, although some receive per capita or fixed-budget reimbursement.
- A pharmacist is paid a fixed amount for a prescription.

Per Capita Reimbursement. This approach reimburses providers a fixed amount per member per month (pmpm), independent of the amount of services provided, resources used, or whether the member is sick or well. Managed care organizations are rapidly expanding this approach to reimbursement. Here are some examples of per capita reimbursement:

- A primary care physician is paid a fixed amount per month for each person listed under his or her ongoing care, for all services provided by the physician.
- A pathologist is paid a fixed amount per month for each person listed under his or her care.

- A managed care organization or insurance company is paid a fixed amount per month to provide a specified set of services, as determined to be required. The issue here is who determines whether, what, and how many services are required.
- Depending on the capitation agreement, primary care physicians and different specialists may receive separate capitation amounts, or there may be a single capitated payment for all physician services.

Fixed-Budget Reimbursement. This approach provides a fixed budget per month or year for all services provided, independent of the number of people served or the amount of services provided. This approach has been most common for health and social services provided by publicly supported agencies. Examples of fixed-budget reimbursement are as follows:

- Many child welfare agencies receive an annual budget from the county.
- Police departments receive an annual budget from the state, county, or city.
- Public health nurse agencies receive an annual budget for nursing services.
- Departments of health and human services receive an annual budget.

Financial Incentives

The financial incentives under these different reimbursement systems are radically different. One major issue is that the incentives for community residents, insured families, managed care organizations, payers, businesses, and different service providers are commonly conflicting. The pressures on and behaviors of the entire healthcare, health, and social services system can be understood by some simple illustrations and descriptions of these incentives.

Although altruism is certainly a factor in the healthcare system, the financial incentives—to maximize the personal and organizational revenues and minimize the personal and organizational costs—are important. Payers, for example, try to minimize their costs and move care to services and settings for which they have no contractual obligation. Similarly, providers try to provide services that maximize their revenues. We will illustrate these concepts with figures that show the general incentives facing traditional hospitals and clinics.

Figure 1.3 illustrates the general financial incentives to use different types of services. In each case, the arrow indicates the dominant direction of the financial incentive. (The upward and downward arrows are shown in different shades to distinguish them visually.) For example, an upward-pointing arrow beside Admissions under fee-for-service reimbursement indicates that the financial incentive is to increase the number of admissions.

FIGURE 1.3. FINANCIAL INCENTIVES
UNDER DIFFERENT REIMBURSEMENT SYSTEMS.

	Fee for Service	Fixed Price per Episode	Per Capita	Fixed Budget
Admissions	↑	↑	↓	↓
Length of stay	↑	↓	↓	↓
Ancillary services	↑	↓	↓	↓
Cost per case	↑	↓	↓	↓
Clinic visits	↑	↑	↓	↓
Skill mix	↑	↓	↓	↓
Withhold services	↓	Mixed	↑	↑
Health	Not Measured	Not Measured	↑	↑

Under fee-for-service reimbursement, the financial incentives are to

- Increase admissions, because each admission generates additional revenues.
- Increase length of stay, because charges increase as patients stay longer in hospitals or nursing homes.
- Increase the use of ancillary services and surgical procedures, because each service is billed separately and increases revenue. This leads to the practice of encouraging, or certainly not discouraging, residents to order many different tests and consults to confirm a diagnosis, and then to determine the usefulness of the different tests and consults later.
- Increase the cost per case, because there are revenues associated with most of the services.
- Increase the number of outpatient clinic visits, because each clinic visit generates revenue.
- Increase the skill mix of staff, by using higher-educated and greater-skilled staff, because these staff justify increased charges.

Because the incentives are to do more of everything, there is little concern about withholding services. There are, of course, some risks of performing additional unnecessary tests and procedures. With few exceptions, although there is certainly general concern about health, there are few measurements or actions taken to maintain or enhance health, because there are significant costs associated with these activities, and virtually none of the payers reimburses for these health-related activities.

Under fixed-price-per-episode reimbursement, some of the incentives change. Overall, the incentives are to provide more of the billable episodes but to minimize the costs of each. Following are examples of financial incentives:

- Increase admissions. Medicare and many other payers reimburse hospitals fixed prices per discharge or admission for patients in each DRG, except for limited numbers of outliers. Hence, each additional admission brings in a fixed incremental amount of revenue, independent of the costs to care for that patient. As long as total costs are less than the amount of revenue received for each patient, the incentive is to admit more patients. If reimbursement for selected admissions falls below costs, there will be economic pressure to reduce those selected admissions.
- Reduce the length of stay for each admission as part of the effort to reduce all costs for the inpatient stay.
- Reduce the use of ancillary services and consultations, again to reduce the costs per admission.
- Increase clinic visits and ambulatory surgery cases because they generate incremental revenue.
- Decrease the skill mix of staff to reduce the cost per case.

The issue of withholding needed services now becomes relevant and should be addressed. Although little factual evidence has been produced, some of the recent reductions of registered nurses for inpatient and outpatient care, to reduce skill mix, have been criticized as being harmful to the quality of care. All stakeholders need better data on quality to answer the questions being raised by shifting reimbursement schedules. One of these questions is whether alterations in healthcare services put patients at greater risk and actually increase costs when some of those patients are later admitted in crisis situations. As with fee-for-service reimbursement, little effort is focused on the health of the communities served.

Per capita reimbursement completely changes the incentives. Since the incentives are similar for per capita and fixed-budget reimbursement, for a given population served, the incentives will be discussed together. In sharp contrast to fee-for-service and fixed-price-per-episode reimbursement, the incentives are to

- Decrease admissions. Since the individual provider or provider organization is paid the same amount pmpm, there is a great incentive to avoid the most costly form of care, inpatient hospitalization.
- Develop alternative, lower-cost alternatives to acute inpatient care (for example, subacute units and home care services).
- Use case-based reimbursement for social services.
- Decrease the length of stay.
- Decrease the use of ancillary services.
- Decrease the cost per case.
- Decrease the number of clinic visits. Most organizations under per capita reimbursement have developed nursing programs that respond to telephone calls from patients to provide information, avoid visits to providers if possible, and route them to the most cost-effective care.
- Reduce the skill mix to the extent possible, while ensuring quality care, to reduce payroll costs. The idea is to avoid using highly paid professional staff to perform activities that could be accomplished equally well by lower-paid staff.

Under per capita and fixed-budget reimbursement, there are legitimate concerns that services may be withheld. Yet despite many newspaper articles alleging that reducing staff and skill levels automatically leads to decreased quality, it is not clear to what extent this is occurring. The primary questions about reducing services are whether the services that are provided are necessary and appropriate and whether they make a difference in outcomes. Better data are required around this issue. Maintaining health among the covered population now becomes a serious concern, because this avoids illnesses and more costly physician visits and hospitalizations.

One difference between per capita and fixed-budget reimbursement is that organizations with per capita reimbursement have a financial incentive to increase the number of covered lives, particularly those of higher health status, because each additional covered life generates revenue. Hence, these organizations will seek new covered lives. Organizations with a fixed budget, however, may avoid additional covered lives because of the increased resource requirements. One additional concern with per capita reimbursement is whether payers will purposely try to exclude the less healthy, more costly members of the community. One humorous, although possibly apocryphal, story was that a payer offering per capita health insurance for Medicare patients put the new membership office on the third floor of a building without an elevator. This approach would prevent disabled and unhealthy elderly citizens from obtaining the insurance and thus reduce the likely costs per member. Another story describes a Medicare HMO that markets enrollment through area health and country clubs, a clever way to minimize services to high-risk, high-user populations.

The financial incentives related to selected foci of care are illustrated in Figure 1.4. Note the almost complete switch of incentives between the fee-for-service and fixed-price-per-episode forms of reimbursement and the per capita and fixed-budget forms of reimbursement. The former have greater financial incentives to provide traditional diagnosis and treatment services because they generate the most revenue, whereas the per capita and fixed-budget incentives favor promotion and prevention to avoid costs associated with using healthcare services.

Similarly, financial incentives related to selected settings of services are illustrated in Figure 1.5. The financial incentives under fee-for-service and fixed-price-

FIGURE 1.4. FINANCIAL INCENTIVES RELATED TO SELECTED FOCI.

	Fee for Service	Fixed Price per Episode	Per Capita	Fixed Budget
Promotion	⬇	⬇	⬆	⬆
Prevention	⬇	⬇	⬆	⬆
Diagnosis	⬆	⬆	⬇	⬇
Treatment	⬆	Mixed	⬇	⬇

FIGURE 1.5. FINANCIAL INCENTIVES RELATED TO SELECTED SETTINGS.

	Fee for Service	Fixed Price per Episode	Per Capita	Fixed Budget
Community	⬇	⬇	⬆	⬆
Home	⬇	⬇	⬆	⬆
Ambulatory	⬆	⬆	⬇	⬇
Inpatient	⬆	⬆	⬇	⬇

per-episode reimbursement favor increased use of traditional inpatient and ambulatory settings. The financial incentives under per capita and fixed-budget reimbursement favor increased use of home and community settings to avoid the more costly inpatient and ambulatory facilities.

There is a humorous but thought-provoking one-liner related to fee-for-service and fixed-price-per-episode reimbursement: healthcare providers would all be in deep trouble or bankrupt if there was an outbreak of health.

Push and Pull Models of Forces on Healthcare Resources

Given the incentives created under different reimbursement systems, there have been major forces to change the use of healthcare personnel and resources. Two different models are described here to illustrate these rapidly changing forces on the use of healthcare resources and the consequences of those forces.

Pull Model. We begin with a simplistic model of the impact of reimbursement on the use of healthcare resources over the past thirty to forty years. Under fee-for-service reimbursement, there has been a strong incentive for more specialized and costly resources to be used, because they can generate greater revenue. This model creates a strong upward draw of patients into the healthcare system, with patients pulled, or drawn higher, into specialized services, as illustrated in Figure 1.6. The model is illustrated with human resources, but there is an accompanying use of all types of specialized services, including high-technology equipment, greater use of intensive care and acute care inpatient services, and more specialized medications.

This tremendous pull of patients into more specialized care created an increasing demand for more specialized resources, as illustrated by arrow 1 in Figure 1.6. The result was shortages of hospital inpatient beds from the 1950s through the 1980s in selected geographic areas and shortages of specialist physicians and registered nurses during the 1970s and 1980s. Thus, the financial incentives and focus on technology and highly specialized services resulted in a system in which patients were continually drawn higher and seen by more specialized providers. At the same time, there was a decrease in demand for less specialized staff, as illustrated by arrow 2 in the figure. This led to decreases in the number of aides, orderlies, and licensed practical nurses.

Push Model. With the shift toward fixed price per episode, and especially toward capitated reimbursement, the model for the use of healthcare resources is completely reversing. Instead of placing incentives to pull patients higher into the system, the incentives under capitated reimbursement are to push patients downward

FIGURE 1.6. PULL MODEL UNDER FEE-FOR-SERVICE REIMBURSEMENT.

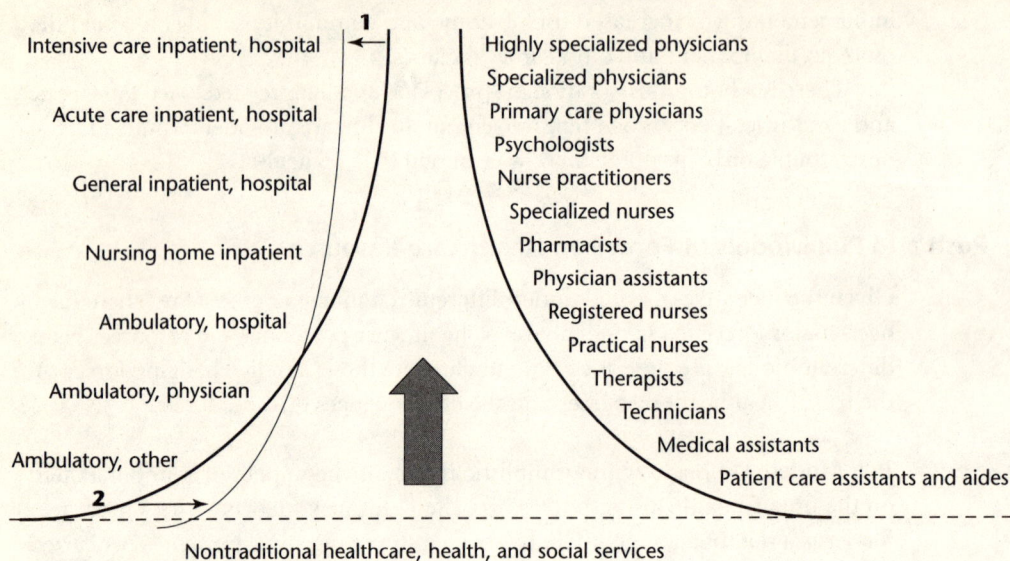

Intensive care inpatient, hospital **1** | Highly specialized physicians

Acute care inpatient, hospital | Specialized physicians

Primary care physicians

General inpatient, hospital | Psychologists

Nurse practitioners

Nursing home inpatient | Specialized nurses

Pharmacists

Ambulatory, hospital | Physician assistants

Registered nurses

Practical nurses

Ambulatory, physician | Therapists

Technicians

Ambulatory, other | Medical assistants

Patient care assistants and aides

2

Nontraditional healthcare, health, and social services

in the system, as illustrated in Figure 1.7. In a simplistic notion, the incentives are now to push patients to lower-skilled providers, using less expensive resources and less costly settings.

The shortages and surpluses of healthcare resources are now the opposite due to the reversed incentives to use less expensive resources. Managed care organizations and other payers are taking many actions to reduce the use of specialized physicians, which is causing a decrease in the demand, as illustrated by arrow 1 in Figure 1.7. Surpluses of specialist physicians, specialist nurses, and registered nurses are now occurring in most metropolitan areas. There is an increasing demand for lower-priced healthcare workers and other resources, as illustrated by arrow 2. In addition, there is a new phenomenon: people not previously considered part of the formal healthcare system—for example, chiropractors, podiatrists, aroma therapists, coaches, and clergy—are now being included, as illustrated by arrow 3. It has been estimated that 33 to 50 percent of all discretionary healthcare dollars are spent on alternative healthcare. As the incentives change from providing more care for the sick to keeping people healthy, whole new classes of people and services become important.

FIGURE 1.7. PUSH MODEL UNDER CAPITATED REIMBURSEMENT.

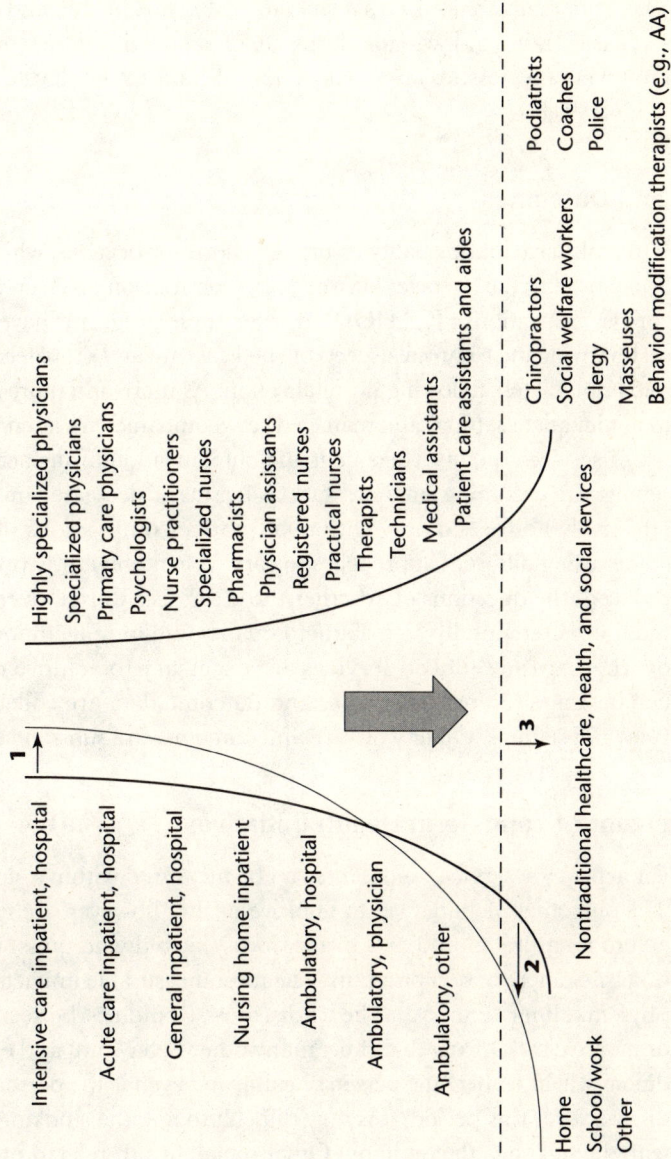

Intensive care inpatient, hospital

Acute care inpatient, hospital

General inpatient, hospital

Nursing home inpatient

Ambulatory, hospital

Ambulatory, physician

Ambulatory, other

Home

School/work

Other

Nontraditional healthcare, health, and social services

Highly specialized physicians

Specialized physicians

Primary care physicians

Psychologists

Nurse practitioners

Specialized nurses

Pharmacists

Physician assistants

Registered nurses

Practical nurses

Therapists

Technicians

Medical assistants

Patient care assistants and aides

Chiropractors

Social welfare workers

Clergy

Masseuses

Behavior modification therapists (e.g., AA)

Podiatrists

Coaches

Police

1

2

3

Strategic Issues Facing Virtually Integrated Health Systems

There are many major issues related to healthcare, health, and social services, and there is certainly no way to address all of them within the context of this book. However, a few issues are particularly relevant to the development and operation of VIHSs.

Quality and Outcomes

Hospitals have made quality assurance efforts for decades, which have been monitored by external agencies like the Joint Commission on Accreditation of Healthcare Organizations (JCAHO). However, these programs have focused primarily on reviewing the technical aspects of medical care and short-term outcomes during and immediately following hospitalization. As more and more services are shifted to noninpatient settings and managed care companies are attempting to restrict the use of services, quality assessment and outcome monitoring across all services and settings are becoming a major issue. Criticisms of decreased quality of care can be effectively addressed only with quantitative comparisons of quality and outcomes using different approaches to care. When previously provided services are reduced, the questions of whether those services were indeed unnecessary and did not incrementally contribute to quality and long-term improved outcomes, or whether the reduced services are necessary to achieve quality outcomes, can be answered only if quality and outcome data are collected across all services and settings. Quality of care and outcomes are substantial issues for VIHSs.

Measurement of Long-Term Health Initiatives

For acute care services, outcomes can be measured within a short period of time. The outcomes of initiatives to improve health, however, are long term and difficult to measure. Consider a twenty-two-year-old who quits smoking as a result of a smoking cessation program. The most measurable impacts of smoking—emphysema, lung cancer, and heart disease—would not be seen for ten to twenty or more years. During that time, many other behavioral and environmental conditions will also affect the person's health; and even if the person manages to avoid diseases over this period, it is very difficult to link the smoking cessation causally with the final health condition. Other social, health-related programs, like working with schools to reduce drinking and driving, the use of drugs, and teenage pregnancies, can be measured in a much shorter period of time because of reduced accidents, injuries, drug overdoses, and teenage births.

There is one interesting economic caveat, although socially questionable. Some economists are finding that the very high-risk activities, especially those with a high likelihood of resulting in death as opposed to long-term disability, may actually have lower lifetime healthcare costs due to the shortened life span. The motto becomes: Die early, save money.

Balance Among Initiatives

With the confusing and conflicting financial incentives, and the long-term impacts of health and social initiatives, the issue is to find a balance of initiatives that help retain an organization's financial viability while working to improve health status. Consider an organization whose services are reimbursed on fee for service or fixed price per episode and whose health and social services are poorly or not at all funded; if it quickly and completely restructures its services toward the wellness paradigm, it will go bankrupt in the short run. If it continues to emphasize billable services and ignores the health risks in the communities it serves and payers switch to per capita reimbursement, it will go bankrupt in the longer term. Per capita reimbursement carries a huge financial risk unless the health of the insured population is maintained, and diseases are treated at an early stage. Let us say too that it is faced with an aging and very unhealthy population, while at the same time receiving less reimbursement for care for these patients. The issue then is to change at the exact pace required, which is extremely hard to accomplish. Changing at a pace to match the environment is very difficult to communicate to physicians and staff because of the apparent conflicting and changing messages. One useful tool to develop an understanding of these environmental and operational dynamics is a simulation program developed by the Healthcare Forum called *Risky Business: Mastering the New Business of Health* (1996).

Population Access Versus Scope of Services

There is a constant tension between the percentage of a population with access to services and the scope and amount of services available (see Figure 1.8). Although the specifics vary substantially, there are really two different approaches to the size of the population covered and scope of services. Virtually all societies provide some basic services to at least part of their populations, illustrated by the darker square in Figure 1.8. Several countries, including Canada, the United Kingdom, and Germany, provide basic healthcare services for all members of their population, which extends the box horizontally to more people. In the United States, the portion of the population that is insured has access to virtually all proved medical technologies, which extends the box vertically. No major country in the world

FIGURE 1.8. PEOPLE COVERED VERSUS SCOPE OF SERVICES.

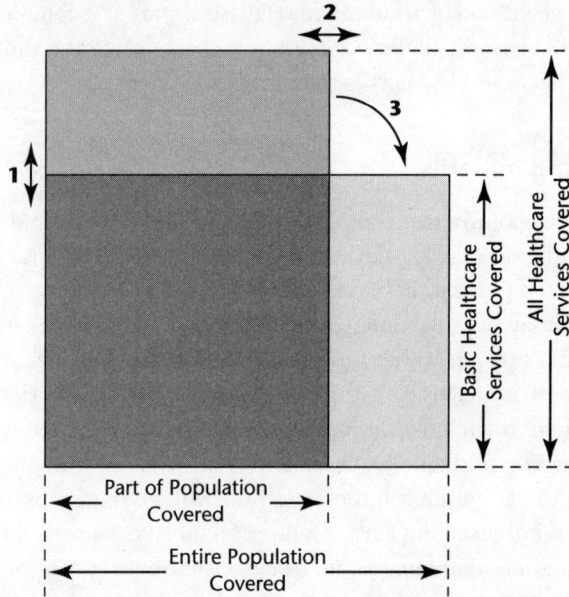

has been willing and able to provide broad availability of all medical technologies to all members of its society. This situation leads to three tensions. Countries that cover the entire population have a continuing tension over the scope of services covered, indicated by 1 on Figure 1.8. People want access to more services, but it is economically and politically impossible for governments to finance this level of service. In the United States, there is a continuing tension over how many people can be eligible for all medical technologies and who will pay for those services, indicated by 2 on Figure 1.8. An additional tension in the United States relates to a reduction of services available to the insured population to finance basic services for additional people, indicated by 3 on Figure 1.8. One of the reasons that federal and state agencies, and major payers, are pressing healthcare providers to reduce costs is to continue to provide similar coverage to currently covered populations and add basic services for more of the uninsured, without exceeding the politically acceptable ability to finance the changes.

Access to healthcare, health, and social services is a growing issue in the United States because the number of uninsured people continues to increase. The U.S. Bureau of the Census estimated that in 1995, 40.6 million people, or 15.4 percent of the total population, lacked health insurance (U.S. Department of Commerce, 1996, p. 1). Since current insurance focuses on acute medical care services only, the number of people without access to appropriate nursing home care, prescription and nonprescription medication, and health and social services is much larger. In addition, a number of other factors are contributing to increasing numbers of uninsured people:

- The increasing costs of health insurance make it more difficult for individuals and organizations to afford. Thus individuals and organizations decrease or altogether drop their health insurance coverage.
- Businesses are dropping health insurance on retirees or requiring increased cost sharing. The healthcare costs for retirees are rising at a disproportionate rate, along with skyrocketing pharmaceutical costs. Major employers are attempting to structure plans and incentives to provide efficient, effective care. This issue has been increasingly salient because of high insurance costs and changes in accounting rules that require businesses to show their financial obligations for future health insurance commitments on their balance sheets.
- There has been growth in the number of small businesses, which typically have small profits and cannot afford the high costs of health insurance. In fact, most employment growth within the United States is within small businesses In 1995, 32.3 percent of people working in businesses with over a thousand employees did not have employment-based health insurance. For smaller organizations, the percentage of people without insurance is much higher: 46.3 percent for companies of twenty-five to ninety-nine employees and 71.7 percent for those with under twenty-five employees (U.S. Department of Commerce, 1996, p. 3).
- Part-time employment has grown. In 1995, 22.4 percent of those who worked part time did not have health insurance, compared to 16.4 percent for those who worked full time (U.S. Department of Commerce, 1996, p. 2).
- Medicare is predicted to go broke shortly after the turn of the century. Even if Congress continues the program, which is likely, benefits are likely to decrease and more of the financial obligation will be transferred to the elderly. In addition, the age of eligibility will rise.

An industrial and operations engineering team at the University of Michigan used different mathematical models to predict the number and percentage of the total population without health insurance. If current trends continue, based on their predictions, the number of uninsured people will reach approximately 62.8 million by the year 2010, as shown in Figure 1.9.

We can anticipate that the federal and state governments, and VIHSs, will have to address the issue of the uninsured. In addition, if we consider the broader scope of health and social services, the problem is even larger.

FIGURE 1.9.   NUMBER AND PERCENTAGE UNINSURED, BY YEAR.

Year	Number of Uninsured (millions)	Percentage Uninsured
1995	40.6	15.4
2000	47.5	17.3
2005	54.9	19.2
2010	62.8	21.1

Source: Alvarez-Buylla, Hamilton, and Korkowski (1996, p. 1).

Churning Among Insurers

Further complicating the strategy for managed care organizations and VIHSs is the fact that people frequently switch insurance plans. A VIHS or insurer that promotes health and social services to improve the health of its covered population may experience a percentage of its insured population switching to other insurers. Thus, the investment made in health and social services may be lost, a phenomenon known as *churning* in the insurance industry. This situation leads to a short-term attitude among some insurers of minimizing investment in health and delaying care as long as possible, knowing that a percentage of the people will switch to other payers in the interim. Clearly, in the longer term, health diminishes and costs increase for all payers. This is an area where legislation may be appropriate to ensure that all payers cover basic health promotion activities, to improve long-term health, and reduce costs, for the whole population.

Information Systems

One obvious issue is the need for much better information systems. In the past, most healthcare information systems have been based on paper medical records and structured to facilitate billing in a fee-for-service reimbursement system in addition to managing medical care. Information systems of the future should:

- Cover the full range of healthcare, health, and social services across a broad range of settings and include the unique interests and priorities of each person.
- Provide continuity of care and health management for patients and their families, with a greater emphasis on health status and risk management.

- Focus on the management of healthcare and health rather than on billing. Clearly, reimbursement is a necessary component, but it should not serve as the primary basis for structuring the information systems.
- Include much more information about outcomes and functionality than is currently the case. Outcome and functionality information are vital for measuring the outcomes of different medical and health management protocols to evaluate their relative quality and cost-effectiveness.
- Include patient and other customer feedback information.
- Move from paper to computerized records. As medical care and health management become more decentralized and involve greater numbers of providers, movement of physical records becomes totally infeasible. Security of information to ensure that only authorized people use the record is a major concern.
- Provide people with ready access to at least major portions of their medical and health record.
- Provide access to people's personal and family records through the Internet, provided that confidentiality issues can be resolved.

Clearly, VIHSs will need greatly improved information systems to manage medical and healthcare in the most cost-effective manner.

Social Issues

As VIHSs take a broader view of the communities they serve, a number of social issues become important. When physicians and hospitals focus only on treating illnesses and injuries in hospitals, clinics, and their offices, there is little immediate concern for social issues. However, under per capita and fixed-budget reimbursement systems, any social issues that affect health become relevant both clinically and financially. Examples of social issues that should be addressed by VIHSs include the following:

- Terminal care and the patient's right to die, including the issue of people with terminal diseases choosing to end their lives.
- Quality of life versus just extending physical life.
- Individual rights versus public rights and public health.
- Crime and violence.
- Accident prevention.
- Substance abuse while driving, working, and engaging in other activities.
- Appropriateness of healthcare, health, and social services and who determines the appropriateness. Balancing the social good with the costs will be an issue.

Alignment of Goals and Incentives

There are currently major differences in goals and incentives among individuals, the community as a whole, insurers, public agencies, and healthcare providers. To reduce conflicts among its members and covered persons, VIHSs should work to increase the alignment among goals and incentives of all stakeholders. Gaucher and Coffey (1997) and Young and Coffey (1997) provide expanded discussions on the concepts and measurements of alignment. The research demonstrates a strong correlation between quality and alignment of goals and behaviors.

Actions for Change

In the remainder of this book, we describe approaches for organizations and VIHSs to address the issues we have presented. First, we set out a taxonomy of healthcare, health, and social services to help in addressing the scope of services, settings, and other dimensions. Then we discuss the use of external expectations and criteria as a basis for assessments of the organization or network. Finally, we present and discuss the vitally important internal organizational climate considerations. We examine analysis tools for each and end with a combined analysis tool to integrate all the information.

CHAPTER TWO

HOW TO ANALYZE VIRTUALLY INTEGRATED HEALTH SYSTEMS AND NETWORKS

This chapter introduces the general analysis approach that we use throughout this book to analyze potential or existing virtually integrated health systems (VIHSs) or networks. Some of the important considerations related to the analyses are also introduced in this chapter.

The analysis provides a mechanism and criteria for a system or organization to assess its strengths, weaknesses, voids, duplications, conflicts, and overall appropriateness. The information it provides then allows the organization to identify and prioritize the issues to be addressed and opportunities to change and expand its business, which in turn improves its ability to meet the needs of the populations served and improve its competitive advantage.

For this book we will operationally define the analyses to include the following assessments:

- Gaps that exist between the needs or expectations and the actual performance (often referred to as gap analyses)
- Duplications of functions or services
- Conflicts among the organizations related to scope of services, external expectations, and internal environmental climates, such as organizational philosophies and practices
- Appropriateness in relation to the needs of the organization, the population being served, and external assessment organizations

Brief interpretations of gaps, duplications, conflicts, and appropriateness are given in Figure 2.1. Each is described in relation to the scope of services, external expectations, and the internal organizational climate.

Types of Analyses

In this book, we focus on three major types of analyses for assessing your own organization and other organizations that may be potential partners or competitors (see Figure 2.2):

1. *Scope of services.* The scope of services and functions provided by your organization and other organizations being considered as potential partners or competitors. A taxonomy of healthcare, health, and social services is offered that describes six dimensions: social and environmental conditions, health-related human conditions, foci, settings, core/key processes, and resources. (The taxonomy is introduced in Chapter Three and serves as the basis describing the scope of services.)

FIGURE 2.1. EXAMPLES OF ANALYSES.

	Scope of Services	External Expectations	Internal Organizational Climate
Gaps	Identify missing services and functions compared to taxonomy	Identify gaps or short-comings compared to expectations	Identify gaps, mis-matches, and critical differences in human dimensions
Duplication	Identify duplications of services and functions	Identify overlapping and duplicate expectations	Identify duplication of human functions
Conflicts	Identify conflicting approaches and services	Identify conflicts among external expectations	Identify conflicts in values, leadership styles, and cultures
Appropriateness	Assess amount and distribution of services appropriate to needs of organization and population served	Assess integrated internal and external expectations appro-priate to meet vision and goals	Assess compatibility and appropriateness of human systems to meet vision and goals

FIGURE 2.2. TYPES OF ANALYSES.

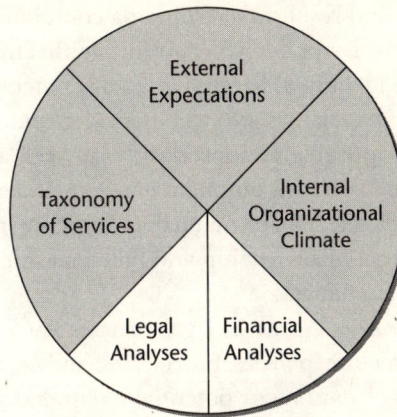

2. *External expectations.* Expectations established by external legal, regulatory, accreditation, and professional organizations. (Criteria from several different organizations are described in Chapter Five.)
3. *Internal organizational climate.* The leadership and characteristics of the internal organizational climate within each organization being considered as part of the system. (Examples of related criteria are described in Chapter Seven.)

Additional analyses, related specifically to quality, financial practices, and legal issues and practices, are also appropriate. Since most networks, and their advisers, have focused the majority of their attention on the financial and legal analyses, we will provide only brief descriptions of these.

• *Quality Analyses.* Separate analyses of quality are done by many organizations, often inspired by the Malcolm Baldrige National Quality Award (Baldrige) criteria to assess organizational excellence. The Baldrige criteria by themselves are an important tool for self-assessment. In Chapters Five and Six, the Baldrige criteria are integrated with other external expectations to develop an inclusive set of criteria.

• *Financial Analyses.* Financial audits, performed by accounting firms, verify that acceptable accounting practices are followed and the financial situation is accurately represented. When forming health systems, several additional financial analyses are included. (We use the term *analysis* rather than *audit*, which has a negative connotation.) The financial analyses are expected to represent the current financial situations fully and accurately. Related to gaps, the financial analysis should identify missing, inappropriate, and conflicting charts of accounts, financial measures, and results. Developing a common chart of accounts, with common definitions, is normally required. Duplications of financial and business systems will be identified. Related to appropriateness, the analyses will assess the consistency of financial systems with the vision and goals of the organization, consistency with applicable standards and practices, and consistency and compatibility of financial systems among partner organizations. The financial analysis may also include selection of manual and computerized systems to be used by the system, resolution of ownership and financial situations, and financing of ownership and other changes.

• *Legal Analyses.* Normally legal analyses assess the appropriateness of the organization's structure, policies, procedures, and practices from a legal perspective. Legal analyses of existing or potential integrated systems involve additional activities. Related to gaps, the legal analyses should address the missing or new legal issues. For example, when a not-for-profit organization merges part or all of its operations with a for-profit organization, issues arise around the disposition of assets contributed by the community over time. Related to duplication, the analyses should address duplicate and conflicting legal structures, policies, and procedures. Related to appropriateness, the consistency of legal policies among participants must be assessed. In addition, when organizations form networks, there are several legal assessments and work related to the ownership, structure changes, and future policies and procedures.

The three major types of analyses can be illustrated as a model of the analyses, shown in Figure 2.3. The completed taxonomy describes the configuration of healthcare, health, and social services and the services provided. Clearly, each health system may choose different configurations to meet the needs of its constituents. The external expectations and human environment describe how the system will accomplish its goals through processes and people. The external expectations are consolidated into a set of common process expectations of external accreditation and professional organizations. The human environment describes a set of cultural and human dimensions associated with successful organizations.

Scope of Analyses

Analyses may be approached with different scopes of comparison and different levels of detail, depending on the goals of the analyses.

FIGURE 2.3. MODEL OF ANALYSES.

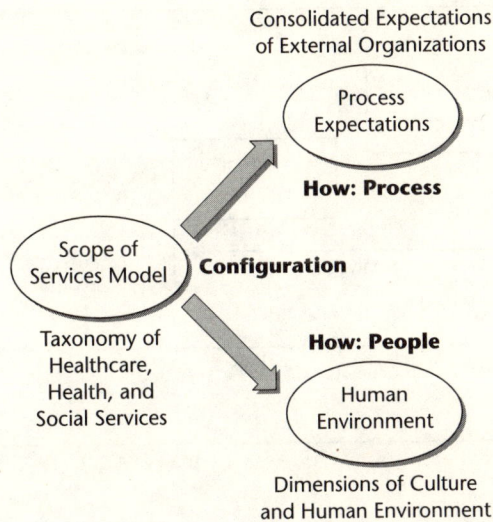

Consolidated Expectations
of External Organizations

Process
Expectations

How: Process

Scope of
Services Model   **Configuration**

Taxonomy of
Healthcare,
Health, and   **How: People**
Social Services
                                Human
                                Environment

Dimensions of Culture
and Human Environment

Scopes of Comparison

Four different scopes of comparison for analyses are illustrated in Figure 2.4. Your organization or system normally begins with a self-assessment (Analysis 1) to gain a better understanding of your own situation and your strengths and weaknesses as part of a virtually integrated health system or network. These analyses can be performed by making internal comparisons to the criteria in this book or best performers within your organization or system. The analyses can also be performed by using comparison data from the units of your organization or other organizations, a process called internal and external benchmarking.

After you understand your own organization, you are ready to make meaningful comparisons with other organizations. Thus, in Analysis 2, you assess your organization and current network partners. This should be done before progressing very far with negotiations with other potential network partners. Analysis 3 examines your organization, your current partners, and potential partners. These analyses are important during negotiations to determine the gaps, duplications, and issues facing the potential partnership. Analyses involving competitors (Analysis 4) should be done on a routine basis. Although Analysis 4 is shown in the figure as being compared to all current and potential partners, it should be done whether your organization is a solo organization or in an existing or potential integrated network. Although your organization can analyze potential partners,

FIGURE 2.4. SCOPE OF COMPARISON AMONG DIFFERENT ORGANIZATIONS.

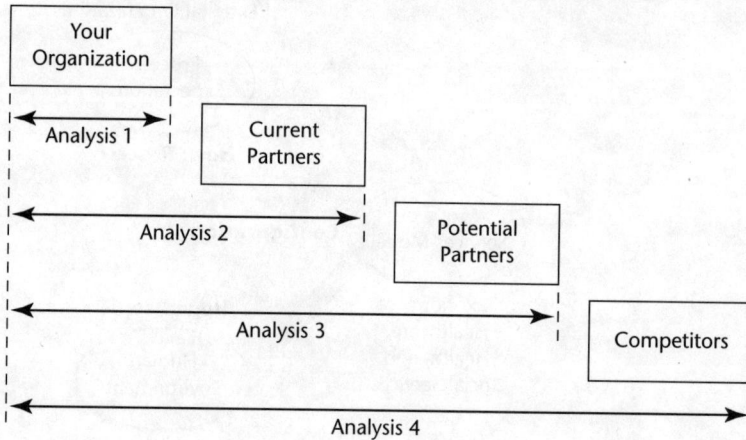

it is best to conduct the analyses jointly, because the analyses will be more complete and accurate.

Level of Detail

Each of these analyses can be performed at different levels of detail. Although the terms do not have common definitions, you normally begin with an overview analysis and then progress to greater levels of detail.

Overview Assessment. An overview assessment looks for major gaps, duplications, conflicts, and issues. The analyses may look at the general services offered but not specifically at how they are provided. The overview assessment is helpful to identify the major issues first, and to prioritize topics for further analysis, before committing large amounts of resources for detailed analyses. Duplicated services and conflicting values cause much greater barriers than gaps, because they require resolution among people with vested interests in their current situations, and potential losses of services and resources. For example, if both your organization and a potential partner organization have strong programs in children's services and cancer care, the appropriateness of duplication, leadership, and philosophies must be resolved. A religious organization may have specific prohibitions against providing certain services, whereas a public institution might be legally required to provide those same services, which could cause a major conflict over values as integration proceeds.

Intermediate-Level Analyses. The intermediate analyses address the services with increased specificity. The intermediate analyses of inpatient care, for example, might determine the volume and cost of diagnosis-related groups (DRGs). On an outpatient basis, you might start looking at the types of services available in different geographic locations. This level of analysis might also address the major types of staff or skill mix used to provide services. The philosophies of care for different social, health, and healthcare services are addressed as well.

Detailed Analyses. The time-consuming and costly detailed analyses are undertaken after major issues are resolved. There is no point in spending huge amounts of time and money analyzing implementation until the major issues are resolved. If those issues are resolved and the partnership moves forward, detailed analyses will ultimately have to be done for most services. Even a simple merger of outpatient services may require detailed changes like selecting a single patient record number, identification and merger of patient records (each facility may have separate records), determination of the contents and order of the information in the medical record, processes to provide continuity of care among the system components, and consolidation of billing systems. Detailed analyses include consolidation of medical and financial records, development of work processes, determination of job classes and responsibilities, resolution of salary and benefit differences, and mechanisms to distribute and refer patients and families for services.

Additional examples will be described in subsequent chapters.

Staffing the Analyses

Depending on the intended aim of the analyses and the depth and level of objectivity desired, the following types of people may be involved. In fact, combinations of people from these different groups may be best:

1. *Internal staff.* Using internal staff to conduct at least the initial analyses is the least expensive and disruptive to the organizations involved, unless they are pulled off other important activities. By participating in the analyses, leaders and staff are sensitized to the strengths, weaknesses, and opportunities of the organization or network, and you gain their understanding and support of the

changes. The weakness of using internal reviewers, unless very specific assessment criteria are used, is that their biases about the organization may affect the objectivity of the assessment.

2. *People internal to the organization but outside the component or division being analyzed.* A common approach, particularly among manufacturing organizations doing quality analyses, is to use people from other divisions of the organization. They tend to be more objective yet have many of the advantages of internal staff.

3. *Members of all organizations involved in the analyses.* Particularly while assessing a potential integrated system, the involved organizations may participate jointly in assessing the scope of services, external expectations, and internal organizational climate.

4. *External reviewers.* External reviewers—consulting companies, accounting firms, legal offices, and others—provide an objective assessment of your organization. One useful approach is to use internal staff first to identify areas that could benefit by more objective external analyses.

5. *Official reviewers.* The external organizations that accredit or review healthcare or social services organizations send reviewers to audit your organization. We define these as official reviewers because they work for the external organization and their purpose is to conduct an assessment for that organization. These official reviewers can also be important sources of information.

The scores obtained with the assessment tools will change with the level of detail of the analyses and the people doing the analyses. An analogy related to the Malcolm Baldrige National Quality Award assessments will be described to illustrate this point. The first overview assessment based on leaders' perceptions using the Baldrige criteria often show a respectable score. However, as more detailed assessments are done later using specific Baldrige criteria, the scores commonly go down, although there have been steady efforts to improve quality. The decline of scores is normally due to the specificity of criteria and an assessment of the degree of deployment, not any real decline in quality.

Uses of Analyses

The analyses described in this and subsequent chapters have different uses. The following are some of the major uses:

- To identify issues and conflicts for resolution. Early identification of key issues is one of the most important benefits of the analyses, particularly related to the internal environmental climate. This allows early resolution of the issues, or alternatively, an early decision not to pursue the system relationship.

- To identify gaps and deficiencies. Whether you pursue a VIHS or not, knowing the gaps and deficiencies of the services and resources will help your organization improve the services it provides.
- To identify opportunities. Some of the identified gaps will be opportunities for new business ventures. This is particularly true for healthcare, health, and social services outside the traditional scope of services.
- To identify duplications, especially if they are inappropriate. Duplications are assessed, whether considering integration of services with existing or potential partners or assessing competitors. However, the potential actions are different. With partners, the goal is to have duplications of services only when appropriate to meet the population needs. With competitors, minimizing inappropriate duplication may be possible but more difficult. In some cases, it becomes a decision to offer duplicate services for continuity of care or access to your health system.
- To assess the compatibility of potential partners in a VIHS. The basic compatibility of potential partners must be assessed as soon as possible to avoid unnecessary effort and expenditures. Many planned mergers have blown up over leadership and other issues after a year or more of intensive, detailed analyses, because underlying issues had not been addressed. During this period, millions of dollars have been wasted.
- To identify and prioritize areas for more detailed analyses. The time and cost requirements to complete detailed analyses are substantial. Completing the overview analyses included in this book will help focus the detailed analyses on the topics most important to making decisions regarding the network. During implementation, virtually all processes must be addressed in detail.
- To identify unique strengths of your organization or health system for use in marketing to potential system partners, payers, businesses, or the public.

Conducting an objective analysis of your organization alone, compared to existing or potential partners or to competitors, is a relatively fast and effective way of identifying the issues to be addressed. In addition to the general analysis approaches described in this chapter, specific analysis criteria will be discussed in subsequent chapters. Several considerations and cautions related to the approach and process of assessment are discussed in Chapter Ten.

CHAPTER THREE

A MODEL FOR UNDERSTANDING VIRTUALLY INTEGRATED HEALTH SYSTEMS

This chapter describes a taxonomy, or classification model, that is useful for describing and planning virtually integrated healthcare, health, and social services. It is intended to expand the thinking of the component organizations beyond traditional medical care to encompass an integrated system aimed at improved health status and quality of life, not just treating acute illness. The taxonomy, which has its origins in a document prepared for Health and Welfare Canada (1979), will assist leaders of healthcare organizations, social services, communities, and businesses in assessing the gaps, overlaps, appropriateness, and opportunities for improving the health of residents within a defined service area. This VIHS taxonomy provides a mechanism for describing the current and desired configuration of needs, services, and resources.

This chapter focuses on the taxonomy portion of the overall assessment process, as illustrated in Figure 3.1.

Characteristics of a Classification Model

A good taxonomy should have the following characteristics:

- *Independence of major classifications.* Although true independence is difficult to achieve, a classification model should avoid situations where the intersec-

FIGURE 3.1. TAXONOMY OF SERVICES.

tions are clearly interdependent. Age and height, for example, are two independent ways of classifying people. The six dimensions of the classification model described in this chapter can also be thought of as a six-dimensional matrix, with the intersections of the classification categories as cells within the matrix.

- *All-inclusiveness.* Each classification category should be subdivided such that all elements relating to that category will be included in the classification model. All elements should be included in the model, even if they are not currently being addressed by your healthcare system, so you will have a complete picture of the needs and opportunities.

- *Consistency.* The classifications used and the measurements taken should be consistent, such that multiple observations of the same phenomenon will yield the same classification. Similarly, operational definitions or interpretations of the classification categories should be as consistent as possible among organizations.

- *Intuitive logic.* Although it may sound obvious, a classification model will be most useful if it has some intuitive logic. For example, the divisions should relate to the way people think and to existing or logical measurement systems.

- *Relation to other classification systems.* A classification model will be easier to understand if it is related to other existing classification systems. Thus, people will be able to translate from one system to another and understand how the systems relate.

General Description of the Taxonomy

The VIHS taxonomy has six basic dimensions or classifications (see Figure 3.2):

1. *Social and environmental conditions.* The health of a population is heavily influenced by the social and environmental conditions of the people living in that area. This dimension, inadequately addressed by traditional healthcare providers, seeks to classify a population by selected social and environmental conditions that are strongly correlated with health and sickness.
2. *Health-related human conditions.* These conditions categorize the physiological, psychological, or health situation of the patient or population being considered. Several different systems are available to classify health-related human conditions. Although there are certainly inadequacies with the International Classification of Diseases, Ninth Revision, Clinical Modification (ICD–9–CM), it is the most comprehensive and broadly used classification system for human conditions. Hence, it will be used to illustrate the approach. You may want to supplement the ICD–9–CM codes with more specific codes, particularly related to specialized services and for health and social services.
3. *Foci.* The categorization of services provided to patients and members of the public is based on the type of interaction between the patient or population and the healthcare or social system.
4. *Settings.* This refers to the geographic location or area where services or activities are provided.

FIGURE 3.2. DIMENSIONS OF THE VIRTUALLY INTEGRATED HEALTH SYSTEM TAXONOMY.

1	2	3	4	5	6
Social and Environmental Conditions	Health-Related Human Conditions	Foci	Settings	Core/Key Processes	Resources

5. *Core/key processes.* These are key processes of any organization, or virtually integrated system of organizations, addressing healthcare, health, and social services.
6. *Resources.* These are the resources—physical, human, and financial—used to provide health-related services and activities.

The categories and level of detail within each of the classification dimensions may be expanded or changed to meet the requirements of the organizations involved.

These classifications provide an excellent framework to assess an existing or potential VIHS and will be used to illustrate the analyses related to the scope of services in Chapter Four. Your organization may choose to supplement or change the model, based on your special circumstances. *Resist eliminating categories from the model simply because your organization currently does not address them.* If your organization or system eliminates all the dimensions and categories of the taxonomy that you do not currently address, the resulting model will fail to highlight additional opportunities. For example, eliminating the home setting category from the taxonomy because you currently do not offer home care services would have the effect of eliminating these services from consideration in the future. One of the primary aims of the taxonomy is to cause you to challenge and expand your current framework. In an industry that has been built around physician offices and hospitals, a tool like the VIHS taxonomy can expand thinking.

A health-focused program brings together a specific set of foci, settings, core/key processes, and resources into an integrated system directed toward a particular human condition within a social and environmental situation. The purpose of any program is to improve the health status of a population or the quality of remaining life. Health status and functionality should be the ultimate measures of success. Also, it is important to realize that healthcare, health, and social programs will vary over time and in response to emergency or disaster situations within the communities served.

Social and Environmental Conditions

The health of a population is strongly correlated with social and environmental conditions. To improve health, a health system must address social and environmental conditions that may negatively affect health. For a number of reasons, including lack of reimbursement, traditional hospitals and other healthcare providers have done little to measure or change social and environmental conditions that adversely affect health. Social service and governmental agencies are well aware of the issues but are often underfunded to address them. Successfully addressing these issues requires coordination among all the stakeholders in the healthcare system and community. However, as the focus of reimbursement shifts and the percentage of people covered by managed care organizations increases, it becomes economically important to focus much more attention on keeping people healthy. That new focus will cause a sharp increase in the amount of attention paid to social and environmental conditions that adversely affect health. Although there are changing incentives, attitudes, and activities toward health, experiences in Arizona, California, and Minnesota suggest that the behaviors start shifting quickly as the percentage of the population covered by per capita insurance exceeds 20 percent. A sample classification of social and environmental conditions is illustrated in Figure 3.3. Additional space is provided at the bottom of the figure to allow additional social and environmental conditions, if appropriate for your situation.

Environmental Pollution

Pollution of water, air, ground, and other aspects of the environment can cause serious health problems for people living in polluted areas. Initially, the key environmental factors affecting health should be identified. The Environmental Protection Agency (EPA) and the state department of natural resources can provide measures of key environmental health risks, if any. These can be used to identify the appropriate types, priorities, and geographic distribution of services. The VIHS can then work with local industries to demonstrate the improvements in health and reductions of costs associated with reducing environmental health risks.

**FIGURE 3.3. CLASSIFICATION OF SOCIAL
AND ENVIRONMENTAL CONDITIONS.**

1
Social and Environmental Conditions
• Environmental pollution • Crime and violence • Community and social support • Family and living situation • Educational and vocational levels • Employment and income levels • Risk factors and behaviors • Political, cultural, and economic environment • Other _____

Crime and Violence

We cannot expect to improve community health when crime and violence are prevalent. In addition, the amount and type of crime and violence will affect the types of health and healthcare services needed by the population. Crime and violence affect not only the people directly killed or injured, but also radically affect the mental health, stability, and attitudes of everyone in the area. In areas with gang activity and violence, most children have experienced the loss or injury of one or more friends. These experiences can cause permanent detrimental mental attitudes and outlooks on life. Hopelessness and fear can take away the desire to do well and improve or escape the current situation.

The aim of classifying areas by degrees of crime and violence is to determine the types and priorities of community, health, and healthcare services. To begin, it is not necessary to have highly detailed analyses of all different crimes. We suggest working with the police and other community agencies to identify which communities or neighborhoods are at high risk for crime and violence. The data for health-related crime and violence can be aggregated into a weighted index to help determine priorities for collaborative efforts of the VIHS and other community groups to reduce the crime and violence most negatively affecting health and mortality.

Community and Social Support

The existence of community and social support mechanisms can significantly affect the physical, mental, social, and spiritual health of the population. The availability of recreational and educational facilities provides positive alternatives for children and other residents. People at these facilities serve as positive role models and lead experiences that help children develop improved abilities and attitudes. Community programs and facilities, such as houses of worship, schools, the YMCA, scouts, libraries, and senior citizen groups, may offer support and development opportunities that can affect health. The measures of community and social support include the number and lists of such programs; the percentage and number of the population, by age group, participating in the programs or using those facilities; and the percentage and number of the population whose health could benefit from joining those groups.

Family and Living Situation

Family and living situations—single-parent families, poor family structure, high-density living situations, elderly people living alone, and difficult living situations of disabled people—can have a direct impact on health. If high-density living situations that may foster communicable diseases are known, then members in the home and community can be prophylactically treated to reduce the spread of disease. Single-parent families have difficulty caring for sick children, which can lead to more days off work and more days of missed school. Elderly people living alone may require living and healthcare support to remain independent and avoid high-cost nursing home care. Proper living situations for disabled and developmentally impaired people can reduce injuries, care requirements, and the associated healthcare costs. The U.S. Census and county and state agencies are sources of these demographic data that may help in targeting risky situations for improvement.

Educational and Vocational Levels

The levels of educational and vocational training of the residents may have an effect on health. Educational level affects the type of work the community residents will do and is a key contributor to income level, which affects the percentage of people with health insurance. In addition, education exposes people to greater knowledge about health. U.S. Census data can be used to determine the income level by geographic area.

Employment and Income Levels

Employment and income level may have several direct and indirect effects on health, including level of access to healthcare and existence of health insurance. U.S. Census data can be used to determine employment by industry, average income, and income distribution, by geographic area served. Occupational Safety and Health Administration (OSHA) data can provide information regarding employment-related health risks. Local employer information can be used to identify employment by employer, health insurance coverage of employees, and the health insurance plans available to those employees. There is an important cycle related to employer-provided health insurance. For example, if the employer purchases health insurance benefits that pay only for emergency and acute care, the employees have financial incentives to use emergency services rather than pay out of pocket for a less costly physician office visit, which then may lead to more days off work and school.

Risk Factors and Behaviors

All of us are aware of behaviors that increase the risk of cancer, heart disease, acquired immune deficiency syndrome (AIDS), and many other diseases. Similarly, alcohol and drug abuse are behaviors that cause major health risks. Many of these risks can be eliminated or reduced. The prevalence of these behaviors allows us to predict the type and amount of services required. More important in the long run is providing effective information and support for people to reduce or eliminate their risky health behaviors. Examples are television ads discouraging illicit drug use and unprotected sex, and providing telephone numbers for further information about the risks.

Political, Cultural, and Economic Environment

The political, cultural, and economic environment within communities may establish limits and opportunities within which a health system must function. For example, if the regional culture and political environment will not tolerate educational programs in the schools, houses of worship, and community settings related to drugs or sex, the health system may have to limit public and school programs to address these important topics affecting the health of your communities.

Health-Related Human Conditions

In order for a healthcare, health, or social program to have an aim, it should be related to the identification, intervention, alteration, prevention, or maintenance of a specific health-related condition. A VIHS can use statistics on health-related conditions to understand the needs of the population and to project the needs for different types of services. An example of a uniform and widely used system of classifying health-related human conditions is the ICD–9–CM. The large majority of hospital inpatients are coded using the ICD–9–CM codes, which are then grouped into diagnosis-related groups (DRGs).

We suggest beginning the assessment of your organization and potential partners using higher levels of aggregation of ICD–9–CM codes. Healthcare, health, and social services programs can be effectively planned for a population by using aggregations of these codes, which are widely used by hospitals and can be aggregated to meet your assessment requirements.

Most of the attention to date has been on classifying various diagnoses, procedures, and reasons for visiting healthcare providers, because that has been the focus of our healthcare system. With the increase in per capita payment for healthcare services, however, there is a greatly increased financial and social interest in the classifications related to health.

For specific services, it may be appropriate to use a greater level of detail or an entirely different coding system tailored to those specific services—for example:

- A program focusing on mental health would expand the classifications of mental disorders and factors influencing health status.
- A program focusing on maintaining and improving health would expand the classifications for services used by healthy people.
- A program focusing on physician services may use coding procedure terminology (CPT) codes for clinical procedures.
- A program focusing on child welfare and family counseling may use direct service codes for those specific services.

The four major categories of classifications within the ICD–9–CM system, plus a classification of maintenance and enhancement of health, are illustrated in Figure 3.4. Space is provided at the bottom of the figure to allow additional classification systems as appropriate, especially related to maintenance of health and family and social welfare services.

Diseases and Injuries

This category provides a subdivision of known and observable diseases or types of injury. The ICD–9–CM classification system (Commission on Professional and Hospital Activities, 1979) provides increasing levels of detail for differentiation. The second level of detail, within diseases and injuries, divides the diseases and injuries into categories by type of problem or body system:

- Infectious and parasitic diseases
- Neoplasms
- Endocrine, nutritional, and metabolic diseases and immunity disorders
- Diseases of the blood and blood-forming organs
- Mental disorders
- Diseases of the nervous system and sense organs
- Diseases of the circulatory system
- Diseases of the respiratory system
- Diseases of the digestive system
- Diseases of the genitourinary system
- Complications of pregnancy, childbirth, and the puerperium
- Diseases of the skin and subcutaneous tissue
- Diseases of the musculoskeletal system and connective tissue
- Congenital anomalies
- Certain conditions originating in the perinatal period
- Symptoms, signs, and ill-defined conditions
- Injury and poisoning

FIGURE 3.4. CLASSIFICATION OF
HEALTH-RELATED HUMAN CONDITIONS.

2
Health-Related Human Conditions
• Diseases and injuries (seventeen major ICD–9–CM categories) • Operations (sixteen major ICD–9–CM categories) • Health status and contact with health services (eight major ICD–9–CM categories) • Causes of injury and poisoning (twenty-two major ICD–9–CM categories) • Maintenance and enhancement of health (classification systems for social services and classification system for health, in addition to ICD–9–CM) • Other _____

Each of these categories is broken into further levels of detail, with explanations for consistent coding.

Operations

The category of operations provides subdivisions of all possible surgical procedures and treatments into basic categories, organized by body system. As with the diseases and injuries, the ICD–9–CM further divides these categories, with explanations for consistent coding (Commission on Professional and Hospital Activities, 1979):

- Operations on the nervous system
- Operations on the endocrine system
- Operations on the eye
- Operations on the ear
- Operations on the nose, mouth, and pharynx
- Operations on the respiratory system
- Operations on the cardiovascular system
- Operations on the hemic and lymphatic system

- Operations on the digestive system
- Operations on the urinary system
- Operations on the male genital organs
- Operations on the female genital organs
- Obstetrical procedures
- Operations on the musculoskeletal system
- Operations on the integumentary system
- Miscellaneous diagnostic and therapeutic procedures

Factors Influencing Health Status and Contact with Health Services

The ICD–9–CM classification system also includes categories for people who are not currently sick but encounter health services for some specific purpose or when a circumstance or problem is present that influences the person's health status but is unknown or is not itself a current illness or injury. Examples addressing potential or existing conditions include prophylactic vaccination, personal problem discussions with a physician, personal or family history of a certain disease, or actions to prevent problems or deterioration related to conditions such as an implant. Examples addressing the maintenance of wellness include weight control, cessation of unhealthy behaviors such as smoking, and exercise programs.

The following ICD–9–CM classifications of factors influencing health status and contact with health services are primarily focused on intervention and are not very specific about health maintenance activities (Commission on Professional and Hospital Activities, 1979):

- Persons with potential health hazards related to communicable diseases
- Persons with potential health hazards related to personal and family history, such as hypertension
- Persons encountering health services in circumstances related to reproduction and development
- Healthy, liveborn infants according to type of birth
- Persons with a condition influencing their health status, such as mental disability
- Persons encountering health services for specific procedures and aftercare
- Persons encountering health services in other circumstances
- Persons without reported diagnosis encountered during examination and investigation of individuals and population (includes routine general medical exams other than for infants and children for prevention and early detection of illness)

External Causes of Injury and Poisoning

These categories include environmental events, circumstances, and conditions that are the potential cause of injury, poisoning, and other adverse effects—for example, transport accidents, accidental poisoning, misadventures to people during surgical and medical care, abnormal reactions and complications, accidental falls, accidents caused by fire, and accidents and potential injury due to natural and environmental factors. The key ICD–9–CM classifications of causes of injury and poisoning are (Commission on Professional and Hospital Activities, 1979):

- Railway accidents
- Motor vehicle traffic accidents
- Motor vehicle nontraffic accidents
- Other road vehicle accidents
- Water transport accidents
- Air and space transport accidents
- Vehicle accidents not elsewhere classifiable
- Accidental poisoning by drugs, medicinal substances, and biologicals
- Accidental poisoning by other solid and liquid substances, gases, and vapors
- Misadventures to patients during surgical and medical care
- Surgical and medical procedures as the cause of abnormal reaction of patient or later complication, without mention of misadventure at time of procedure
- Accidental falls
- Accidents caused by fire and flames
- Accidents due to natural and environmental factors
- Accidents caused by submersion, suffocation, and foreign bodies
- Other accidents
- Late effects of accidental injury
- Drugs, medicinal, and biological substances causing adverse effects in therapeutic use
- Suicide and self-inflicted injury
- Homicide and injury purposely inflicted by other persons
- Legal intervention
- Injury undetermined, whether accidentally or purposely inflicted

All of these health-related human conditions are more precisely defined by the third- and lower-order subclassifications.

Maintenance and Enhancement of Health

The ICD–9–CM and other classification systems used to describe medical and healthcare services oriented toward curative medical care are inadequate for the expanded view of maintaining and enhancing health. Hence, classification of health-related efforts needs improvement. A second, difficult issue related to maintaining and enhancing health is the breadth of factors involved. Several such factors were described in Figure 1.1 on the paradigm shift from sickness to wellness. Some key factors are:

- *People.* As we consider health-oriented services, many additional types of people are involved, such as exercise physiologists, dietary consultants, exercise consultants, hypnotists, and clergy. Many of these people are not licensed or recognized by traditional healthcare providers. (See Figure 1.7.)
- *Settings.* The settings are expanded to include community centers, houses of worship, recreational fields, public highways, and other locations where people exercise, play sports, meditate, and meet with others who form their social support network.
- *Equipment.* The equipment to maintain and enhance health can encompass sports equipment, home exercise equipment, and theoretically even books used to improve mental status.
- *Finances.* The costs associated with maintaining and enhancing health could include many of the things individuals and organizations do, which is much broader and harder to measure than the costs of hospitals, nursing homes, and other licensed healthcare providers.

Sets of Health-Related Human Conditions

It is possible to analyze each of the health-related human conditions defined by the ICD–9–CM classification system, plus other classification systems tailored to your particular VIHS. However, we have found that it is more useful to aggregate the health-related human conditions into sets that are related to the population served and programmatic competencies to address the population needs. The level of detail of aggregation should be based on the level of analyses being performed. For example, aggregations of ICD–9–CM codes related to diseases of the circulatory system and operations on the cardiovascular system may serve as a good basis for analyzing the degree to which you and your VIHS partners address heart disease to serve your population. Analyzing each related ICD–9–CM code is too detailed and wastes time, effort, and money at this state. Later, for the detailed analyses—for example, when you want to analyze which cardiac procedures will be done in which of your facilities—the more elaborate ICD–9–CM codes would be appropriate. In the detailed-analysis stage, other detailed coding systems, such as CPT codes, may also be appropriate.

Foci

Health and healthcare services can be grouped into foci, or categories, based on the focus of the services provided or interaction with the patient or population member. The fourteen foci illustrated in Figure 3.5 cover the range of foci related to healthcare, health, and social systems. As is clear, this covers a broader range of services than normally considered by hospitals and other healthcare organizations. Space is provided for additional foci, if considered appropriate for your VIHS.

Planning activities related to a given focus are considered part of that focus. For example, planning the treatment and rehabilitative care for a psychiatric patient is included in the respective treatment and rehabilitation foci.

This section describes fourteen broad foci of healthcare, health, and social systems, with second-order classifications and examples. Although some of the

FIGURE 3.5. CLASSIFICATION OF FOCI.

3
Foci
• Promotion • Protection • Prevention • Detection • Diagnosis and assessment • Treatment • Habilitation and rehabilitation • Maintenance • Hospice • Support • Advocacy • Education • Research • Enabling • Other _____

examples refer to federal programs, states and provinces, counties, and cities may have similar or additional programs.

Promotion

Health promotion foci are oriented toward promoting health through healthy behaviors and attitudes. The aim is to keep people healthy, so they do not use traditional healthcare services. Sample health promotion foci are given here as illustrations.

Health Education Focus. These are services directed toward informing, educating, and motivating the public to adopt personal lifestyles and nutritional practices designed to promote optimal health, avoid health risks, and make appropriate use of healthcare and social services in the community. Health education services include transfer of health knowledge, transfer of health information, and motivation toward positive health behavior (including the modification of poor health habits). Labels on food that provide information about fat content, calories, and other nutritional facts enable people to make informed choices. Media public service ads educate people about places to call for different kinds of assistance.

Health Promotion Focus. These are services designed to promote optimum well-being among individuals in the community. Similar activities may accomplish both health education and health promotion. This is most effective when done collaboratively with the families, schools, employers, houses of worship, police, the media, and other community entities. Examples include programs to address drug abuse, alcohol abuse, tobacco smoking, and family conflict resolution. Examples of a health promotion focus are television and radio commercials encouraging parents to get their children immunized and cautioning against fallen power lines. These are particularly effective when local, regional, and national role models are communicating the message.

Protection

Protection foci are actions or interventions oriented to protecting people from known and unknown health hazards in the community, workplace, school, home, or other public places. As with promotion, the aim of protection is to keep people healthy. Collaboration of efforts is important among the VIHS and many other organizations working on protection efforts. Sample protection efforts follow.

Environmental Quality Management. These are measures taken to protect the community from environmental hazards causing or contributing to disease, illness, injury, or death. Environmental hazards include air, water, and noise pollution, as well as hazards related specifically to unsafe residential and community environs. Prominent concerns of environmental quality management include water supply treatment and wastewater disposal; solid and hazardous waste disposal; air pollution control; noise control; housing and residential hazards control; vector control; recreational area hazards control; highway safety; and assessment of the health effects of environmental pollutants; assessment and control of health hazards and effectiveness of medical devices, radiation-emitting devices, and hazardous products; and assessment of health effects of technological and sociological environments.

Food Protection. These are measures taken to ensure wholesome and clean food free from unsafe bacteria and chemical contamination, natural or added hazardous substances, and decomposition during production, processing, packaging, distribution, storage, preparation, and service; and to ensure that marketed foods comply with established nutritional, quality, and packaging identification guidelines. Sample major components of food protection are sanitation, safety, and nutritional quality, and food quality and hazards, including nutritional content, microbial hazards, and chemical hazards (both added and natural). Truth in advertising is also a consideration in food and other consumer projects.

Occupational Health and Safety. These are actions taken to ensure the recognition, prevention, and control of occupational health hazards and illness and to promote the physical and mental well-being of employed persons. Occupational health and safety concerns are typically different for different occupational categories (for example, mining, construction, agriculture, transportation, utilities, manufacturing, services, food services, and health services). Examples are installing safety guards on equipment and following uniform body substance precautions when caring for patients, such as wearing gloves when drawing blood.

Radiation Safety. These are actions taken to protect the community from unnecessary exposure to ionizing and nonionizing radiation from controllable industrial and nuclear sources and to minimize exposure of patients and medical personnel to clinical radiation. Sample major areas of concern in radiation safety are industrial radiation, medical radiation, and radioactive wastes.

Biomedical and Consumer Product Safety. These are measures taken to ensure that drugs, cosmetics, therapeutic devices, and all other types of consumer products, including cleaning fluids, pesticides, and children's toys, are safe and appropriate for their intended use and are clearly labeled as to the potential harm resulting from abuse or misuse. Sample major categories of biomedical and consumer product safety include the following:

- Drug quality and hazards, including assessment of effectiveness and wise use of drugs, microbial, and chemical hazards in the drug and cosmetic supplies and control of the movement of narcotic, and other drugs subject to abuse.
- Health surveillance, including provision of national, state, and local health and disease information; the provision of a national reference service for identification of disease-producing bacteria, viruses, and parasites; and the assessment and improvement of laboratory diagnostic procedures.
- Public safety, including guidelines and surveillance of public transportation safety, public building safety, and safety in public parks, streets, and other public places. Examples related to public transportation include licensing pilots, air crew, and air traffic controllers; licensing drivers of public ground transporters; medical assessment of pilots; and flight safety and accident prevention programs. Police and fire protection are also public safety activities. Testing of water at public beaches is another form of public safety activity.
- Personal protection, such as wearing seat belts in automobiles, wearing helmets on motorcycles, and wearing protective gloves, boots, and glasses when working. Several hospitals are now giving infant car seats to parents of newborn babies. Similarly, Health Partners of Southern Arizona in Tucson and the Children's Hospital Medical Center in Cincinnati give bicycle helmets to children of their insured families and patients to protect them from head injuries.

Greatly increased international travel has added to the danger of communicable diseases, plant disease, and pests being transferred among countries. Therefore, public health efforts to control communicable diseases are also important protection efforts.

Prevention

Prevention foci are actions taken specifically to prevent illness or accidents. Depending on the organization, there may be some overlap between protection and prevention activities. Personal examples of prevention include immunization, dental prophylaxis, mental health consultation, family counseling, and health maintenance activities.

Many organizations offer employee assistance programs to employees and family members to address health and family problems resulting from stress, substance abuse, or other problems. Early counseling and family assistance, for example, can prevent abuse of children, spouses, and others. Most programs ensure confidentially, so the employee or family member can continue his or her job.

Detection

Detection services identify or measure physical, mental, chemical, social, and other characteristics of individuals in presymptomatic, unrecognized symptomatic, or symptomatic stages of human conditions. The aim of detection is information gathering, which serves as input to the diagnosis and assessment foci. Alternatively, some organizations include the measurements of a person with symptoms, along with the integration of information, as part of the diagnosis and assessment foci.

Detection activities can be described in terms of the physical, mental, chemical, or other measurements being made. Alternatively, detection can be classified in terms of the technology used to make the detection. Sample second-order classification is described here in terms of the characteristic of the individual that is measured:

- Physical measurements, such as height, weight, size, and strength
- Visual measurements, such as acuity
- Auditory measurements, such as sound thresholds
- Tactile measurements, such as sensation of pain or muscle reactions
- Olfactory measurements, such as the ability to detect different odors
- Gustatory or taste measurements, such as the ability to taste various chemicals
- Respiratory measurement, such as respiratory rate or oxygen uptake
- Biochemical measurements, such as blood sugar or cholesterol level
- Imaging measurements to determine the existence, size, or condition of internal organs and masses (for example, using X rays, ultrasound, and computerized tomography scans)
- Internal observations and measurements, such as endoscopies or blood pressure
- Tissue measurements of excised tissue samples or organs
- Mental measurements, such as tests for intelligence or depression
- Functional status assessment, including functioning of families
- Detection or surveillance for accidents

Diagnosis and Assessment

Diagnosis is the process of integrating several types of information to make a judgment about the level of an individual's health or illness. Sample types of information integrated into the judgment follow:

- Measurements from the detection focus. There may be several such measurements. Some include measurements related to symptomatic conditions within the diagnosis focus
- The individual's history of health-related events and information, such as family history, medical records, immunization history, dental records, and current exercise program.
- Statistical information about similar sets of data relating to other patients or populations. For example, 70 percent of all patients with a particular measurement were found to have diabetes.
- Diagnostic algorithms or critical paths (Coffey and others, 1992; Coffey, Othman, and Walters, 1995; Coffey, Richards, Wintermeyer-Pingel, and Le Roy, 1996).
- An individual's perception of his or her own situation, well-being, and symptoms.

Information from the different sources can be integrated by one or more of the following methods to reach a diagnosis:

- *Human integration and decision making.* This is done by a health or healthcare professional, such as a physician, nurse, social worker, family care worker, trained technician, or adviser.
- *Machine integration and decision making.* With the expanded complexity of information, computers are being used to integrate information to make judgments or diagnoses. Content experts develop the decision-making methodologies but are not present when judgments are made for individual situations. Complete machine-made judgments are uncommon.
- *Combined human and machine integration of information and decision making.* The combined systems are the most common, with professionals using the information from computer systems to make the final diagnosis. A major issue related to management of information is wide access to information while maintaining patient and family confidentiality. A common use of combined decision making is using machines to alert healthcare professionals when certain situations occur. An example is a cardiac monitor, which sounds an alarm and begins recording information when particular cardiac arrhythmias occur.

Treatment

Treatment refers to actions taken in an attempt to alleviate disease, ill health, or symptoms and to maintain health. Treatment normally is based on a diagnosis, which is based on detection information. However, it is common for detection, diagnosis, and treatment to occur interactively at the same time.

A sample second-order classification of the treatment focus can be based on the different treatment modes, including the following:

- Medication, such as oral medication, intravenous, injection, and respiratory
- Physical treatments, such as setting a broken bone or returning a joint to its correct position, and physical and occupational therapy
- Surgery, such as removing a diseased organ, repairing damaged tissue, or repairing a degenerated or damaged joint
- Radiation therapy
- Prosthetics and orthotics, such as replacement or assistance of body components by specially constructed devices, such as dental bridges, prosthetic limbs, or support devices
- Behavioral therapy, such as crisis intervention and psychological counseling
- Life support, such as cardiac resuscitation and renal dialysis (acute)
- Lifestyle changes, such as changed diet, smoking cessation, and exercise
- Family and environmental changes
- Education of patient, family, and others regarding the treatment and expected outcomes

Habilitation and Rehabilitation

This focus involves activities that help ill, disabled, or physically or developmentally impaired individuals achieve the fullest physical, mental, social, vocational, and economic usefulness of which they are capable. Habilitation involves developing new capabilities, and rehabilitation involves restoring capabilities that have been lost or impaired—for example:

- *Medical habilitation and rehabilitation services.* The medical needs of the ill or disabled individual are evaluated, and a habilitation or rehabilitation device or service to meet the person's needs is designed, prepared, managed, and evaluated.
- *Therapy services.* Therapeutic techniques are used in implementing a program of habilitation or rehabilitation designed to meet the needs of an ill, disabled, or impaired individual. Illustrative therapies are physical therapy, occupational

therapy, recreation therapy, social therapy, audiology, physical education (as for geriatric patients), arts and crafts, and psychiatric rehabilitation services.

- *Assistive devices.* These devices assist the patient and family to achieve the highest degree of functionality. Assistive devices include prosthetic and orthotic devices and services; hearing aids; wheelchairs and carts; crutches, canes, and walkers; and home revisions.
- *Education and training.* These programs teach or demonstrate how to improve one's own functional capability.

Maintenance

Maintenance encompasses services to individuals with chronic conditions in order to prevent deterioration in those conditions or to maintain the maximum level of functioning. This is contrasted with the treatment and habilitation and rehabilitation foci that concentrate on improving the condition. Sample maintenance services include continuing medication, periodic treatment or therapy (such as renal dialysis), continual nursing care and monitoring, assistance with activities of daily living, assistive devices, dietetic services, transportation, social services, and companionship.

Hospice

Hospice services provide important support to the patient and family of people in the last, deteriorating stage of their lives. All parties understand that death is close and that the objective is to keep the patient comfortable, maintain as much quality and dignity of life as possible, and assist the rest of the family with the transition. A hospice typically coordinates counseling, nursing care, equipment, and supplies.

Support

The support foci are services that facilitate the provision and management of medical, healthcare, health, or social services—for example:

- *Admitting and registration.* The admitting and registration process is responsible for scheduling, arranging, and processing an inpatient admission, outpatient service, or service in other setting. The process collects and communicates necessary information to a wide variety of stakeholders.
- *Biomedical engineering.* These activities are responsible for applying engineering concepts and methods to the design and construction of instruments, pros-

thetics, and systems used in patient care. They also provide consulting, maintenance, and repair services for complex medical electronic instruments and equipment.

- *Barber and hairdresser for patients.*
- *Central supply.* These activities are responsible for storing, controlling, processing, sterilizing, assembling, and distributing medical surgical supplies and equipment.
- *Food and nutrition services.* Food refers to the procurement, production, and distribution of food to meet the needs and requirements of patients or other people involved. Nutrition services refer to the clinical assessment and recommendation of diets to support the clinical care and improve the person's health.
- *Health records.* This is a particularly problematic support focus, as health and healthcare services are geographically distributed.
- *Insurance and financial services.*
- *Materiel services.* These provide medical, surgical, and office supplies and related items and services.
- *Pharmacy services.* Similar to food and nutrition services, there are important clinical foci in addition to providing pharmaceuticals.
- *Purchasing.*
- *Administration and management.* Although this group leads the efforts, administration is considered support because its purpose is to plan, direct, coordinate, and control resources to meet the service requirements of the patient or population involved.

Advocacy

The advocacy focus involves the active communication and promotion of the interests of a patient, family, or group of people. Advocacy activities are typically aimed at the people or organizations that can make changes benefiting the affected person or group—for example, those who sponsor and promote the Special Olympics for mentally disabled people or those who lobby for easy wheelchair access to parking, sidewalks, and buildings. Child welfare and family service agencies have pursued client advocacy as a core competency and traditional service.

Education

The education focus encompasses activities that promote and accomplish the broad dissemination of information and transfer of knowledge. This focus is directed at patients, families, and the public and the people who provide services to them. Education of health and healthcare professionals such as physicians, nurses, technicians, social workers, caseworkers, exercise coaches, and administrators is also part of this focus. Sample educational approaches include formal courses, lectures, clinical conferences, internships, residencies, apprenticeships, case studies, correspondence, mentoring, and Internet communications.

Research

The research focus refers to activities that apply the scientific approach to investigation or inquiry, with the intention of improving the health and healthcare system. Sample classifications of research include the following:

- *Clinical research:* investigations into the biomedical or behavioral aspects of health, healthcare, and human systems, using the scientific method
- *Technological or basic science research:* scientific investigation of the performance of health and healthcare systems that aims to advance the level of technology
- *Organizational research:* studies into the composition and structure of health and healthcare systems, directed toward developing organizational theory and improved organizations
- *Services delivery research:* research that strives to define new or improved approaches to the process of delivering health and healthcare services

Enabling

These are organized activities designed to influence the means by which, and the conditions under which, health and healthcare services are provided—for example:

- *Health planning:* a process that establishes desired future levels for health status and healthcare system performance, designs and selects among alternative actions aimed at modifying the health system so that future levels of both health status and health system performance conform to the desired levels, and suggests steps for implementation of the recommended actions. Sample components of health planning are data assembly and analysis, goal determination, action recommendation, and implementation strategy.
- *Evaluation:* analysis or measurement of historical or current situations, to assess changes designed to improve the situation.
- *Financing:* the sources and methods of financing, including both resources development and health system services. Areas of concern include the sources, amounts, and appropriate use of financing from businesses, third-party reimbursement, direct public expenditures for services, public grants, philanthropic grants and payments for service, direct payment by those receiving services, and availability of loans and loan guarantees.

- *Regulation:* the intervention of government or accreditation organizations in the health system by means of rules and regulations that typically influence, control, or set standards relative to the services provided within the health system, the settings in which services are provided, and the manner in which providers of services are reimbursed. Sample classifications of regulation include personnel licensure and certification, facility licensure and certification, equipment licensure and certification, methods used, information collected and reported, and reimbursement methods and rates.

Settings

Several basic types of settings are applicable to the healthcare and health system (see Figure 3.6). These cover the full range of locations in which healthcare, health, and social services activities take place, many of which are outside the typical settings served by hospitals or other healthcare organizations.

Eleven categories of healthcare, health, and social service settings are illustrated in Figure 3.6. Space is provided at the bottom for additional settings, if considered appropriate for your organization.

FIGURE 3.6. CLASSIFICATION OF SETTINGS.

4
Settings
• Area wide • Community • Houses of worship • Schools • Work • Mobile • Home, including assisted living • Ambulatory • Partial-day care • Inpatient • Freestanding support • Other _____

Areawide Setting

The areawide setting covers the whole area or community, often through some form of public communication. Services provided or considered in this setting are typically health protection and health system enabling in nature. Sample areawide health efforts include television, radio, and billboard programming and ads related to the adverse affects of drugs, drinking and driving, AIDS, and child abuse.

Community Setting

Realigning the healthcare system to serve the community will place greater emphasis on operating within community settings rather than within typical healthcare facilities. A community setting is a facility or event established primarily to serve the public rather than a specific set of healthcare providers. These settings are typically owned by some community agency. Sample community settings are neighborhood health centers, public health centers, mental health centers, alcoholism and drug abuse treatment centers, family planning clinics, rehabilitation centers, suicide prevention centers, senior citizen centers, YMCA and other recreational centers, supervised (sheltered) care houses, family care centers, and community centers. Staff may be regularly scheduled in these settings or come only at specific times.

Houses of Worship

Houses of worship provide a unique opportunity to influence health through a peer group that shares religious beliefs. These are particularly good settings for some health and social services because religious beliefs often provide strong support groups. Some organizations have already recognized this important linkage through establishment of "parish nurse" programs to serve congregations through the use of a healthcare liaison. The clergy have often proved influential in matters related to behavioral and family problems.

Schools

Schools are an important setting for the learning and social development of children. As such, they provide an excellent setting for education and social activities oriented toward promotion, protection, and prevention in order to improve and maintain health. Schools may be the best setting in which to advocate not smoking, avoiding illicit drugs, weight control, balanced diet, and other health-oriented programs. Sample services in the school setting (which includes residential schools, colleges, and universities) are hearing and vision tests, first-aid clinics, school nurses, gymnasiums and exercise programs, healthcare facilities, and health education and safety classes.

Work

Most adults spend more time at work than at any other single activity. In addition, keeping employees healthy and on the job is of direct financial interest to employers. Many employers provide employee assistance programs for their employees, to reduce time off work and improve productivity. Some employers provide time and space for healthcare, health, and social services for their employees at or near the work site. Similar to schools, sample services in the work setting are hearing and vision tests, first-aid clinics, and gymnasiums and exercise programs. Healthcare facilities within residential institutions such as prisons may also be considered a work setting. It is clear that as we focus more on health enhancement and maintenance, workplaces, schools, houses of worship, the community, and other sites will be preferred over traditional healthcare sites.

Mobile

These are movable structures, specially equipped vehicles, or staff and equipment used to provide healthcare, health, and social services in multiple locations. A mobile setting is often used to ensure geographic accessibility for an identified target population. This setting also encompasses vehicles used to provide healthcare services during the transporting of patients, such as ambulances. Example mobile settings are ambulances, helicopters, airplanes, mobile health screening units, mobile CT scanners, mobile renal dialysis units, meals on wheels, mobile immunization units, mobile health education displays, Red Cross mobile blood collection units, hospital ships, and movable military surgical suites. The armed forces have highly developed mobile health systems, because they have to provide healthcare services wherever forces are deployed.

Home

A person's usual or temporary place of residence may be a site for healthcare, health, and social services to reduce the necessity for travel and to reduce costs. The staff, supplies, equipment, and information are brought to the person's home. Examples of services provided in a home setting include visiting nurses, hospice care, home intravenous therapy, home-based equipment, homemaker services, occupational and physical therapy services, therapeutic foster care, family counselors, senior day care, respite care, and patient education.

Ambulatory

This is a location where organized services are provided to people who travel to the site for services, receive services, and then leave the site. There is no provision at this type of setting for overnight stay. Example settings for ambulatory care include the following:

- Hospital site—any hospital-based outpatient clinic, emergency department, or ancillary department serving ambulatory patients or clients.
- Separate facilities serving multiple healthcare providers, such as ambulatory surgery centers and imaging centers.
- Healthcare provider offices, such as clinics and offices of specialty physicians, primary care physicians, dentists, chiropractors, physical therapists, and nutritionists. Most providers considered outside the traditional healthcare system have offices or clinics, such as acupuncture and aroma therapists.
- Family services and counseling centers.
- Health clubs, separate from work or community settings.

Partial-Day Care

A partial-day care setting provides supportive care for a patient or other client in a supervised setting for less than twenty-four hours. This setting, a cross between an ambulatory and inpatient setting, usually does not require beds for patients. The duration and services vary with the requirements of the people receiving the services. Observation beds in hospitals may be considered an inpatient setting or partial-day care setting. Other examples are adult or child day care, partial-day foster care, and a special education center.

Inpatient

This location refers to places where personal healthcare, health, and social services are provided to patients who stay overnight. The facilities, equipment, supplies, and staff are regularly available on-site for patients. Examples are general hospitals, specialty hospitals, subacute care facilities, nursing homes, and hostels. Hostels are inpatient hospice institutions that provide care specifically for patients with terminal illnesses. They focus on keeping the patient comfortable and on the patient's psychological and family needs. Although the facilities are similar to those of nursing homes, the focus is different.

Freestanding Support

Services are provided that support the delivery of health and healthcare services, although the location is not a component of an organization that delivers health and healthcare services. Examples are medical laboratories, dental laboratories, pharmacies, tissue and blood banks, and drug information centers.

Core/Key Processes

There are several core or key processes required within organizations focused on healthcare, health, and social services (see Figure 3.7). These are processes within the organization that cross foci and settings. Given the substantial overlap among these processes in the taxonomy and process expectations of external organizations, these core/key processes will be addressed and expanded in Chapters Five and Six. External organizations like the Joint Commission on Accreditation of Healthcare Organizations (JCAHO), the National Committee for Quality Assurance (NCQA), the Council on Accreditation (COA), the Malcolm Baldrige

FIGURE 3.7. CLASSIFICATION OF CORE/KEY PROCESSES.

5
Core/Key Processes
• Leadership and governance • Strategic planning • Human resources management and development • Process and quality improvement • Information planning and management • Continuum of care • Client/patient rights and satisfaction • Prevention and education • Managing the environment • Other _____

National Quality Award, and state quality awards emphasize these same core processes.

Resources

The resources used in the healthcare system can be divided into five broad categories, as illustrated in Figure 3.8. There are many subcategories within each of these broad categories.

Human Resources

This category includes all of the people associated with healthcare, health, and social services, including people associated with the direct provision of healthcare, various types of support activities, health promotion activities, patients and their families, and healthy people and their families.

Healthcare and Social Professions. These are individuals with formal (academic) training and accreditation or licenses in a specific profession who directly provide healthcare, health, and social services to patients and well people. A role of the VIHS is to integrate many services and organizations within an area into effective networks that address the needs of the population in a more integrated, high-quality, and cost-effective manner. Examples of licensed or registered healthcare professions include the following:

FIGURE 3.8. CLASSIFICATION OF RESOURCES.

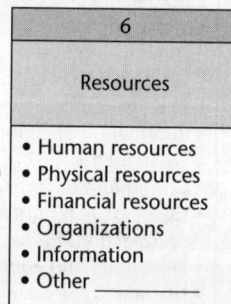

6
Resources
• Human resources • Physical resources • Financial resources • Organizations • Information • Other _____

- Medical doctors and doctors of osteopathy in several categories, including

  General practice, including family practice, general pediatrics, and general obstetrics

  Medical specialties, including dermatology, internal medicine, pediatric specialties, and psychiatry

  Surgical specialties, including general surgery, neurological surgery, obstetrics and gynecology, ophthalmology, orthopedic surgery, otolaryngology, plastic surgery, thoracic surgery, and urology

  Other specialties, including anesthesiology, child psychiatry, pathology, neurology, psychiatry, physical medicine and rehabilitation, radiation oncology, radiology (diagnostic and therapeutic), and miscellaneous other specialties

- Dentists
- Chiropractors
- Optometrists
- Pharmacists
- Podiatrists
- Nurse practitioners, including certified registered nurse anesthetists
- Nurses, including registered nurses and licensed practical nurses
- Midwives
- Physician assistants
- Psychologists
- Physiotherapists
- Audiologists
- Occupational therapists
- Orthotists and prosthetists
- Speech therapists
- Dietitians
- Laboratory technologists
- Radiation technologists, including radiological and radiotherapy technologists
- Other technologists
- Social workers and counselors.

Allied Health Workers. This category refers to persons with formal or informal training or experience who primarily provide services in support of health and healthcare services. The definition and people included in this category may vary from one geographic area to another. Most of these people have some form of formal education, licensure, or certification. Example categories of allied health personnel follow:

- Medical laboratory personnel, including laboratory scientists, laboratory assistants, laboratory technicians, and phlebotomists
- Radiologic technology personnel, such as physicists, radiology technicians, nuclear medicine physicists, nuclear medicine chemists, nuclear medicine technologists, and nuclear medicine assistants
- Dietetic and nutritional personnel
- Rehabilitation personnel, such as audiology technicians, recreational therapists, and recreational assistants
- Other allied medical personnel, such as paramedics, respiratory technologists, renal dialysis technicians, psychological technicians, pharmacy assistants, biomedical engineers, biomedical technicians, child life workers, child life assistants, and cardiology technicians

Depending on the organization and licensure requirements within a geographic area, there may be other classifications, and there may be shifts between health professions and allied health staff.

Assistive Caregivers. Many different types of people provide direct services for patients and clients. Since they are unlicensed, their training and job duties are typically defined by each organization. As networks are being evaluated or formed, these job categories may have substantial differences that must be addressed. Examples of assistive caregiver titles are medical assistants, patient care assistants, nurse aides, orderlies, and home health aides.

Other Supportive Personnel. There is a huge variety of people within healthcare, health, and social services organizations who are important to the maintenance of health, care of patients, and maintenance of an organization but who may or may not provide direct services to patients, clients, and families. As we move away from traditional healthcare services, this distinction becomes difficult. For example, how would you categorize a football coach who trains and develops team members to avoid injuries? The coach works directly with the clients, but they are not considered patients in a traditional healthcare system until an injury occurs. Following are examples of supportive personnel:

- Health record personnel, including registered record analysts, health record analysts, health record technicians, health record librarians, and transcriptionists
- Clergy
- Coaches
- Decision support and financial analysts

- General positions, including secretaries, transcriptionists, accountants, accounting clerks, cashiers, operators, and clerks
- Personnel officers
- Managers and supervisors
- Other support personnel, such as purchasing officers, drivers, laundry equipment operators, librarians, library technicians, volunteers, systems analysts, programmers, computer operators, planners, insurance specialists, and federal and state or provincial employees
- Foster parents, caseworkers, counselors, and mentors

Physical Resources

The physical resources include different types of facilities, equipment, and land used in the healthcare, health, and social services system.

Buildings. This category encompasses the physical structures that house related functions. As we move from specialized healthcare services to health services, health promotion, and family services, the types of facilities involved change dramatically. For specialized healthcare services, there are dedicated facilities such as hospitals. However, for health, health promotion, and family services functions, community, work, school, and home facilities are used on a shared basis.

Equipment. This category refers to all electromechanical or other equipment used in the course of performing healthcare and health functions. In the context of health planning, attention is commonly limited to large, expensive, and highly specialized equipment. Sample categories of equipment include building services equipment, fixed major equipment, portable major equipment, and minor equipment.

As with buildings, as attention shifts to health and health promotion functions, equipment tends to be shared with daily activities in communities, at work, at school, and at home. One example of technology that has tremendous potential to improve healthcare and health throughout the world falls under the generic term of telemedicine. Photographs, sound, and knowledge can be communicated across the world instantaneously. Picture the following scenario that is technologically feasible now.

A trained assistant preps a patient's knee and clamps it in place in a treatment room in India. A young surgeon, using high-resolution video equipment and robotics controlled by satellite communications, performs arthroscopic surgery on that knee, using a video monitor and "joy sticks" to control the equipment from her home in Boston. She is in constant communication with the patient and assistant through the use of video and voice communication. This may sound like science fiction, but it is feasible with today's technology. As we move more toward behavior modification related to health and wellness, communication of knowledge is easier than this surgical procedure. The interpersonal relationship, of course, is best when people are together. However, these technological advances can be vitally important to people in rural and other remote areas lacking trained health professionals.

Land and Land Improvements. Land, parking lots, roads, and so forth are resources necessary to provide healthcare and health services. For specialized healthcare services, there tend to be specific locations owned by healthcare organizations, but for health services, the land and facilities tend to be owned by nonhealthcare organizations and private individuals.

Financial Resources

Financial resources are the sources and amounts of capital and operating funds available to support healthcare, health, and social services—for example, operating income (the net income from services provided) and capital funds (the money used to finance capital purchases, which may come from retained earnings, contributed capital, and debt financing, such as loans, mortgages, and bonds).

Organizations

Although organizations are not normally thought of as resources, they are significant resources. An organization has structure, policies, procedures, and customs to define and accomplish goals. Those same organizations can command other resources. For example, a community organization to coordinate financial resources could be used to assist the coordination of health services. A large industrial company can exercise substantial clout to promote health-related changes in a community.

Information

Information is a valuable resource, especially when combined with information systems to distribute the information broadly to those who have a legitimate need for it, including patients and their authorized family members. These resources are carefully linked to the key process of managing information, discussed in Chapter Five.

Using the Taxonomy

The consolidated VIHS taxonomy is illustrated in Figure 3.9. This can be used as an abbreviated checklist for reviewing VIHSs.

The VIHS taxonomy has several potential uses, including the following:

- Standardizing terminology. An accepted classification model can help standardize the terminology that federal and state agencies, healthcare institutions, social service agencies, payers, healthcare professionals, professional associations, the press, and others use. Because standardization of terminology nationally will be slow, if it ever occurs, your VIHS and the organizations with which you interact can standardize your terminology as you refine the taxonomy for your use.
- Standardizing organization and classification of the healthcare, health, and social services system. Standardization will improve communications and data collection activities among agencies and organizations by providing a common structure.
- Providing a conceptual framework to evaluate current operations and development activities. The classification model will broaden the alternatives considered during the development and evaluation of activities and services.
- Providing a conceptual format to plan and develop a VIHS. One of the important uses of the taxonomy is to stretch your thinking and planning related to what is included in this system.
- Stimulating new services. The taxonomy is a helpful tool to facilitate the discussion of new configurations of healthcare, health, and social services.

Although this taxonomy may need to be revised to meet the needs of your unique organizational or system situation, the taxonomy provides a useful, objective approach to analyzing your current and potential healthcare, health, and social services. A tool to conduct this analysis is discussed in Chapter Four.

Appropriateness Assessment

The taxonomy can be used directly to assess gaps and overlaps of foci, settings, and resources. However, some duplication of foci, settings, and resources is appropriate for a large VIHS. The question is how to determine appropriateness, since it can be judged from different perspectives. From a healthcare and health perspective, we propose that the primary perspective be that of the communities being served. With this perspective, appropriateness is judged from population-based

FIGURE 3.9. VIRTUALLY INTEGRATED HEALTH SYSTEM TAXONOMY.

1 Social and Environmental Conditions	2 Health-Related Human Conditions	3 Foci	4 Settings	5 Core/Key Processes	6 Resources
• Environmental pollution • Crime and violence • Community and social support • Family and living situation • Educational and vocational levels • Employment and income levels • Risk factors and behaviors • Other _____	• Diseases and injuries (seventeen major ICD–9–CM categories) • Operations (sixteen major ICD–9–CM categories) • Health status and contact with health services (eight major ICD–9–CM categories) • Causes of injury and poisoning (twenty-two major ICD–9–CM categories) • Maintenance and enhancement of health (classification systems for social services and classification system for health, in addition to ICD–9–CM) • Other _____	• Promotion • Protection • Prevention • Detection • Diagnosis and assessment • Treatment • Habilitation and rehabilitation • Maintenance • Hospice • Support • Advocacy • Education • Research • Enabling • Other _____	• Area wide • Community • Houses of worship • Schools • Work • Mobile • Home, including assisted living • Ambulatory • Partial-day care • Inpatient • Freestanding support • Other _____	• Leadership and governance • Strategic planning • Human resources management and development • Process and quality improvement • Information planning and management • Continuum of care • Client/patient rights and satisfaction • Prevention and education • Managing the environment • Other _____	• Human resources • Facility resources • Equipment resources • Financial resources • Organizations • Information • Other _____

planning estimates of the most cost-effective combination of foci, settings, and resources to meet the needs of the population. Hence, within the VIHS you would duplicate services only to the extent that it is appropriate to serve your members and communities. However, a VIHS may choose to duplicate services of competitors in order to provide a complete portfolio of services to its members and communities. Education and research have other perspectives of appropriateness, based on their unique requirements.

Following this logic, once the inventory of foci, settings, and resources is determined, population-based estimates of need are required to assess the appropriateness of duplications and the amount of services required in an underserved area. These analyses are complicated because they have the same risk as omitting categories from the VIHS taxonomy. If historically based ratios of resources per 1,000 population are used without considering whole new configurations of healthcare, health, and social services, then you will repeat the same mix of services historically offered. New ratios of facilities and resources will be required, based on the newest, most cost-effective configurations of foci, settings, core/key processes, and resources to meet the population's health needs. One useful approach is to develop computer models to simulate the impacts of different configurations.

Challenges of Change

Using the taxonomy to develop a VIHS to optimize the health of the members and communities served has great potential but also presents great challenges.

Threats to Status. There are many threats to the status of current healthcare providers and organizations, which cause major resistance. Hospitals, specialist physicians, and others providing highly specialized care are economically and professionally threatened. Some of the changes will lead to changes in their social status and challenges to their autonomy. For example, many physicians view critical paths and clinical guidelines as threatening to their professional autonomy, even through these have been demonstrated to improve quality and reduce costs.

Financing Mechanisms. Use of the VIHS taxonomy will identify new types of foci, settings, services, and resources. Yet there are no financing mechanisms for many of these new services, because current designs have been built around traditional medical care approaches. The new approaches will be viable only if there are ways to finance them. Resistance to change will be most easily exercised by simply not financing new services. The impacts of these financing mechanisms are discussed in Chapter Eleven.

Zero-Based Resource Requirements. The most logical approach is to define alternative configurations of foci and settings to address best the social and environmental conditions and health-related human conditions of members and communities and then to do a zero-based development of the core/key processes and resources to meet the needs most cost-effectively. This zero-based development of resource requirements will be resisted and challenged by all groups whose jobs are threatened by any new configuration.

Quality. Quality is the ultimate concern. Like appropriateness, it is a matter of whose perception of what characteristic. Are we talking about quality of technical clinical procedures, quality of service, or quality of life? Are we talking about quality as perceived by the patient, by the provider, or by the payer? A VIHS must address quality as perceived by its different customers and stakeholders.

Leadership. Certainly the process of using the VIHS taxonomy and making decisions based on it will pose several challenges to the leadership of all current and potential VIHS components.

CHAPTER FOUR

ASSESSING THE NEW SCOPE OF SERVICES

This chapter examines the use of a specific assessment tool for the scope of services, based on the VIHS taxonomy described in Chapter Three. Additional analyses related to external expectations and internal organizational climate are described in Chapters Six and Eight, respectively. The aim of the scope-of-services analysis is to inspire your organization and your VIHS partners to broaden your analyses to include activities not traditionally covered. The analysis tools will help clinical and administrative leaders and staff assess whole new configurations of healthcare, health, and social services for your system. A consolidation of the three analyses is described in Chapter Ten, along with a less detailed overall analysis tool. In addition, these tools should help your organization structure and prioritize its assessments so that you can make decisions more quickly. In today's market, the ability to adapt quickly to the environment is crucial.

One important characteristic of a virtually integrated health system is that the whole system address all of the healthcare, health, and social services needs of the populations served. That does not mean that your organization must own or control everything; it simply means that there is a collaborative system to meet all needs in a high-quality, cost-effective manner.

Approach

Before proceeding to detailed analyses, we suggest beginning with an overview assessment that looks for the major gaps, overlaps, and obvious issues that must be addressed. Overview analyses may be conducted by a representative group of administrative and clinical leaders and staff who are familiar with data for the respective organizations. Since the overview analyses are often based more on perceptions than analyses of data, it is important to include people with different perspectives to ensure a complete and balanced view. The second stage is an intermediate level of detail, and should be based more on data than perceptions. Highly detailed analyses are reserved for topics requiring detailed assessment.

The tool presented in this chapter is used to analyze the scope of services. The analyses will include measurements and judgments about the following characteristics:

- Gaps or deficiencies in relation to each component of the taxonomy. These are often referred to as gap analyses. For example, if the VIHS does not actively address health promotion, it would be considered a gap.
- Duplications of functions or services for each dimension of the taxonomy. Duplications are not necessarily good or bad. However, the appropriateness of all the duplications should be assessed, since consolidation of services can often improve quality and reduce costs. Ambulatory services are often duplicated in different settings to provide appropriate geographic distribution of services.
- Conflicts identified during analyses of the scope of services.
- Appropriateness in relation to the needs the population being served and the organization. (Selected questions to address appropriateness will be discussed later in this chapter.)

Different types of people may participate in the scope-of-services analyses, depending on the purposes of the analyses:

- Internal leaders and staff. Especially for the overview assessment of your organization's capabilities, this is the place to start.
- Staff internal to a multifacility organization but outside the division being analyzed.
- The organizations represented in the analyses.
- Representatives of customer groups. Depending on the services your organization provides, these might include patients, families, insured people who have

not used medical services, uninsured community representatives, businesses, students, and referring physicians.

• External reviewers, such as visiting professionals or consultants.
• Official reviewers from the external accrediting, licensing, or certification organizations.

Horizontal and Vertical Approaches

Exhaustive evaluation of all the functions, settings, core/key processes, and resources for every health-related human condition can be unnecessarily and unreasonably time-consuming. More focused analyses are often appropriate, while still using the taxonomy to stretch your thinking. Two different approaches are helpful in using the taxonomy.

Horizontal Analysis

By looking at the taxonomy horizontally, you select one social or environmental condition, or health-related human condition, and then look horizontally at the possible foci, settings, core/key processes, and resources that might apply to that condition. The horizontal arrow in Figure 4.1 illustrates this horizontal approach. Similarly, you could begin with a particular function or setting and then look across at the human conditions, processes, and resources that would be involved. When planning clinical programs, you would use the horizontal analysis approach to address all the foci, settings, resources, and so forth associated with the program for the specific population being addressed.

As an abbreviated example, consider the development or enhancement of a heart program. You would consider all of the current and potential foci, followed by the most cost-effective settings, processes, and resources to provide those services. The promotion focus may include television ads to encourage healthy diets and exercise programs to reduce the risks of heart disease. In large part, these activities would take place in the areawide, community, houses of worship, school, and work settings. The information management process would be included and would require staff to develop the promotion materials, facilities, and equipment to produce the videos and money to cover the costs. At the diagnosis and treatment foci, there may be a combination of outpatient and inpatient services coordinated through the use of care protocols and critical paths. The resources may include cardiologists, thoracic surgeons, nurses, nutritionists, inpatient beds, operating room time, clinic time, and medications. The key is to do an exhaustive assessment of all the social and environmental conditions, foci, settings, core/key

FIGURE 4.1. HORIZONTAL INTEGRATION OF SCOPE OF SERVICES.

1	2	3	4	5	6
Social and Environmental Conditions	Health-Related Human Conditions	Foci	Settings	Core/Key Processes	Resources
• Environmental pollution • Crime and violence • Community and social support • Family and living situation • Educational and vocational levels • Employment and income levels • Risk factors and behaviors • Other ___	• Diseases and injuries (seventeen major ICD–9–CM categories) • Operations (sixteen major ICD–9–CM categories) • Health status and contact with health services (eight major ICD–9–CM categories) • Causes of injury and poisoning (twenty-two major ICD–9–CM categories) • Maintenance and enhancement of health (classification systems for social services and classification system for health, in addition to ICD–9–CM) • Other ___	• Promotion • Protection • Prevention • Detection • Diagnosis and assessment • Treatment • Habilitation and rehabilitation • Maintenance • Hospice • Support • Advocacy • Education • Research • Enabling • Other ___	• Area wide • Community • Houses of worship • Schools • Work • Mobile • Home, including assisted living • Ambulatory • Partial-day care • Inpatient • Freestanding support • Other ___	• Leadership and governance • Strategic planning • Human resources management and development • Process and quality improvement • Information planning and management • Continuum of care • Client/patient rights and satisfaction • Prevention and education • Managing the environment • Other ___	• Human resources • Facility resources • Equipment resources • Financial resources • Organizations • Information • Other ___

processes, and resources related to a heart program. By doing this, you can expand the thinking beyond the current configuration of services and settings.

Vertical Analysis

Another approach is to look vertically at one or more of the six dimensions in the taxonomy to ensure that all the categories within that dimension have been considered. The vertical arrow in Figure 4.2 illustrates the approach of looking vertically at each dimension of the taxonomy. Using this approach related to settings, for example, causes you to think about all the possible settings in which services could be provided and whether you are addressing those settings.

As an example, consider all the possible settings for services that your organization currently provides and plans to provide. We suggest asking the following questions for each setting:

• Is our organization actively involved in providing healthcare, health, and social services in this setting? For example, what are we doing to use areawide communication actively to improve the health of the populations we serve?
• What are all of the known social and environmental conditions, health-related human conditions, foci, core/key processes, and resources potentially involved with this setting? Refer to the literature for unusual services provided in each setting that demonstrate cost-effectiveness and high quality.
• What partnerships and relationships will facilitate improvement of health through this setting? For example, our organization may want to form an alliance with local schools and houses of worship to promote reduced drug use.
• Can we achieve greater improvement in health through other services and settings?

The intent is to cause your organization to think outside its existing foci, settings, and resources.

Analysis Tool

We will use a common format for the analysis tools throughout this book. Clearly, the analysis tools should be tailored to the needs of each organization and the purpose of the analyses. However, just because you are not currently addressing specific aspects of the taxonomy does not mean you should summarily eliminate dimensions or criteria within the taxonomy analysis tool. If you do, you may be setting yourself up for failure to identify potential new services and opportunities.

FIGURE 4.2. VERTICAL INTEGRATION OF SCOPE OF SERVICES.

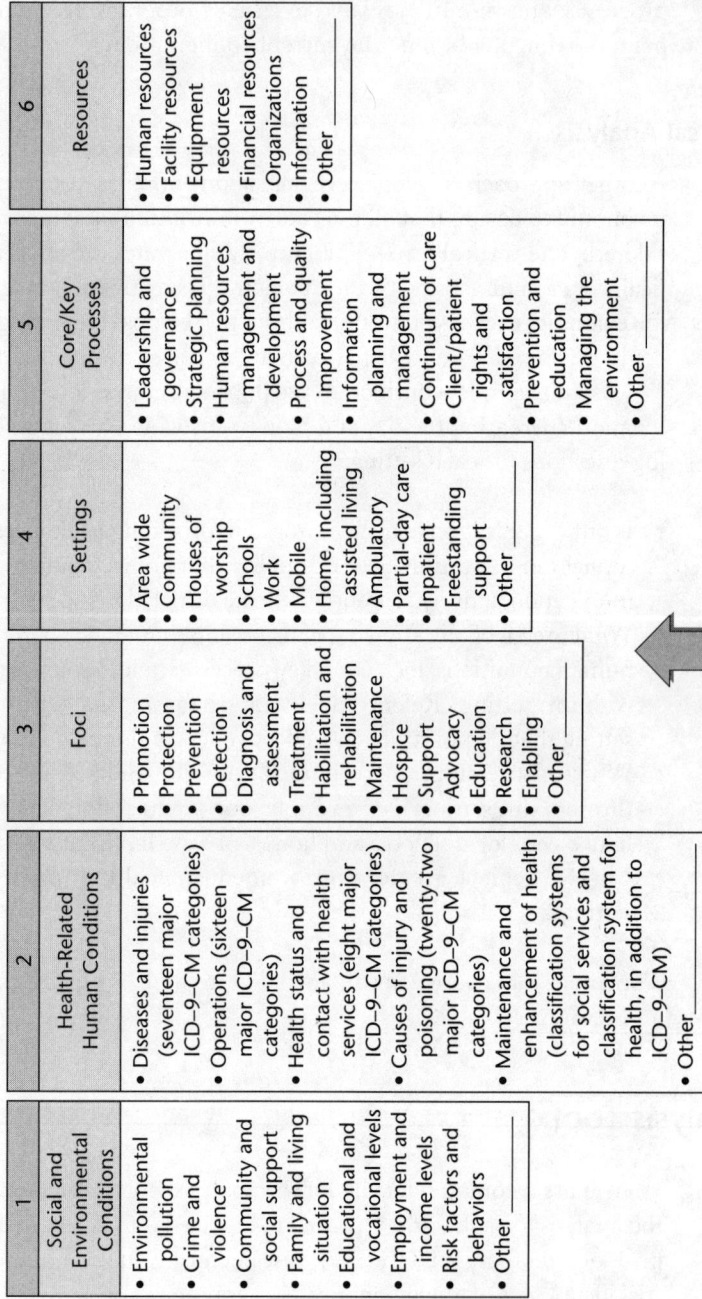

1	2	3	4	5	6
Social and Environmental Conditions	Health-Related Human Conditions	Foci	Settings	Core/Key Processes	Resources
• Environmental pollution • Crime and violence • Community and social support • Family and living situation • Educational and vocational levels • Employment and income levels • Risk factors and behaviors • Other ____	• Diseases and injuries (seventeen major ICD–9–CM categories) • Operations (sixteen major ICD–9–CM categories) • Health status and contact with health services (eight major ICD–9–CM categories) • Causes of injury and poisoning (twenty-two major ICD–9–CM categories) • Maintenance and enhancement of health (classification systems for social services and classification system for health, in addition to ICD–9–CM) • Other ____	• Promotion • Protection • Prevention • Detection • Diagnosis and assessment • Treatment • Habilitation and rehabilitation • Maintenance • Hospice • Support • Advocacy • Education • Research • Enabling • Other ____	• Area wide • Community • Houses of worship • Schools • Work • Mobile • Home, including assisted living • Ambulatory • Partial-day care • Inpatient • Freestanding support • Other ____	• Leadership and governance • Strategic planning • Human resources management and development • Process and quality improvement • Information planning and management • Continuum of care • Client/patient rights and satisfaction • Prevention and education • Managing the environment • Other ____	• Human resources • Facility resources • Equipment resources • Financial resources • Organizations • Information • Other ____

A sample tool for analyzing the scope of services is illustrated in Figure 4.3. Selected key criteria are listed for each of the six dimensions of the taxonomy to analyze the scope of services. At the bottom of each list of criteria is space to add other criteria your organization considers important. You will also note some blank columns in the figure, which will be used later for the combined analysis matrix described in Chapter Ten.

Three columns (d, e, and f) allow scoring of three different organizations related to each of the important criteria: your own organization (column d) and potential or existing VIHS partners or competitors, depending on which organizations are included in the analysis (columns e and f). Of course, additional columns may be added to the spreadsheet for additional organizations. Especially for the initial, general analyses, a simple scoring system is recommended. The question you are trying to answer is how well your organization, its VIHS partners, and competitors are addressing the needs of the marketplace, by analyzing different components of the taxonomy. A five-level scoring system is illustrated in Figure 4.3:

0   No competence, coverage, or success, or not applicable

1   Minimal competence, coverage, or success

2   Partial competence, coverage, or success

3   Significant competence, coverage, or success

4   Substantial competence, coverage, or success

The scores can be assigned from different perspectives. For a VIHS, the scores should be assigned from the perspective of the populations served. (A score of 1 is assigned for minimal effort addressing a given component of the taxonomy.) Using this scoring approach, you are essentially basing your decisions on healthcare, health, and social services needs of the population, which is commonly known as population-based planning.

Since there are likely to be differences among departments or services, you may choose to assign scores based on different approaches:

- A blended score, which considers all of the different units and assigns a score based on a weighted average of all the units.
- A low score. This is the approach used by the Joint Commission on Accreditation of Healthcare Organizations when it assigns a poor score and recommendation for change based on finding poor performance in one area.
- Differences among areas, since this gives an indication of variation within your organization or between your organization and the comparison organizations.

FIGURE 4.3. SCOPE-OF-SERVICES ANALYSIS TOOL.

Major Section		Important Criteria	Criteria Scores Score (0–4, 4 = high)			Comments
Number	Name	Description	Your Organization			
	a	b	c	d	e	f

Scoring of criteria (fill in the boxed areas):

0 = No competence, coverage, or success, or not applicable   3 = Significant competence, coverage, or success
1 = Minimal competence, coverage, or success   4 = Substantial competence, coverage, or success
2 = Partial competence, coverage, or success

1 Scope of Services

Social and Environmental Conditions
Environmental pollution
Crime and violence
Community and social support
Family and living situation
Education and vocational levels
Employment and income levels
Risk factors and behaviors
Other priority conditions,
  identified by organization

Health or Clinical Conditions
Diseases and injuries (seventeen
  categories)
Operations (sixteen categories)
Health status and contact with health
  services (eight categories)
Causes of injury and poisoning
  (twenty-two categories)
Maintenance and enhancement
  of health

FIGURE 4.3. SCOPE-OF-SERVICES ANALYSIS TOOL. (continued)

Major Section		Important Criteria				Criteria Scores Score (0–4, 4 = high) Your Organization			Comments
Number	Name	Description	a	b	c	d	e	f	
		Other priority conditions, identified by organization							
		Foci							
		Promotion							
		Protection							
		Prevention							
		Detection							
		Diagnosis and assessment							
		Treatment							
		Habilitation and rehabilitation							
		Maintenance							
		Hospice							
		Support							
		Advocacy							
		Education							
		Research							
		Enabling							
		Other priority foci, identified by organization							
		Settings							
		Areawide							
		Community							
		Houses of worship							

FIGURE 4.3. SCOPE-OF-SERVICES ANALYSIS TOOL. (continued)

Major Section		Important Criteria	Criteria Scores Score (0–4, 4 = high)			
Number	Name	Description	Your Organization			Comments
	a	b c	d	e	f	

Schools
Work
Mobile
Home
Ambulatory
Partial-day care
Inpatient
Freestanding support
Other priority settings, identified
  by organization

Core/Key Processes (may score in external expectations)

Leadership and governance
Strategic planning
Human resources management
  and development
Process and quality improvement
Information planning and
  management
Continuum of care
Client/patient rights and satisfaction
Prevention and education
Managing the environment

FIGURE 4.3. SCOPE-OF-SERVICES ANALYSIS TOOL. (continued)

Major Section		Important Criteria	Criteria Scores Score (0–4, 4 = high)				Comments
Number	Name	Description	b	c	Your Organization d	e	f	
	a							

Other priority criteria, identified by organization

Resources
Human resources
  Physicians—specialists
  Physicians—generalists
  Physicians—residents
  Physician assistant/
    nurse practitioner
  Nurses
  Pharmacists
  Therapists
  Technicians and allied health
  Medical assistants
  Office staff
  Professional/administrative
  Trades
  Service/maintenance
  Community health nurses/workers
  Child/family welfare workers
  Teachers and coaches
  Clergy
  Influential community members
  Patient/client and family

FIGURE 4.3. SCOPE-OF-SERVICES ANALYSIS TOOL. (continued)

Major Section		Important Criteria		Criteria Scores Score (0–4, 4 = high) Your Organization			Comments
Number	Name	Description					
	a	b	c	d	e	f	
		Other human resources					
		Physical resources					
		Facilities					
		Equipment (particularly expensive/specialized equipment)					
		Instruments					
		Supplies					
		Other physical resources					
		Financial resources					
		Organizations					
		Information					
		Other priority criteria, identified by organization					

A column is included for comments. Although any comments may be helpful, the entries in this column are most helpful if they address the needs for clarification, more detailed analyses, and in particular the type of comparative information desired.

Discussion and consideration of the scores is important. A VIHS should address the large majority of the taxonomy. However, a component organization should not be penalized for a low score in an aspect of the taxonomy outside its core role or competence. If your organization does not cover a specific focus, setting, or resource, that is fine, as long as the populations' needs are covered by another component of the VIHS or another provider.

Analyses of Scope of Services

This section provides example interpretations related to the six dimensions of the taxonomy. The aim of these analyses is to expand your thinking beyond your current services or even the services provided by your partners and competitors.

Social and Environmental Conditions

Seven criteria are illustrated in Figure 4.3: environmental pollution, crime and violence, community and social support, family and living situation, education and vocational levels, employment and income levels, and risk factors and behaviors. Additional criteria can be added, as shown in Figures 4.1 and 4.2, and each of these criteria can be addressed at much greater levels of detail. For the general overview analysis, the following questions can be used to assign a criterion score:

- Does the organization formally address and make an impact on this criterion? For most criteria in the taxonomy, one or more staff members within your organization commit personal time and effort to those initiatives. The question here, however, is whether the organization has any identified, formal program or budgeted effort.
- How much information have you collected and analyzed about the criterion? For example, many healthcare organizations have very few data on or understanding of the amount of driving under the influence of alcohol or drugs, the number of teenage pregnancies, the number of people in the community who smoke, or the number of days people are away from work and school due to illness.
- To what degree has the organization demonstrated success in making an impact on the criterion?

- Are there inappropriate duplications or conflicts among efforts within your VIHS?

In the comments section, it may be helpful to identify the major gaps, duplications, conflicts, and issues, so they can be addressed.

A couple of examples may be useful. Clearly crime and violence have a major impact on the health of your communities. All healthcare organizations respond to the injuries and deaths resulting from crime and violence in emergency departments and clinics. However, those reactive responses are not the intention of addressing social and environmental conditions, and those responses alone should be scored zero. On the other hand, if an organization has identified the sets of risky behaviors most prevalent in its communities and established cooperative programs with houses of worship and schools that have showed a slow decline in those risky behaviors, then it may be scored 2 or 3.

Health or Clinical Conditions

The five criteria illustrated in Figure 4.3, from the ICD–9–CM (Commission on Professional and Hospital Activities, 1979), provide a reasonable classification of diseases, operations, injuries and poisoning, and selected health status conditions. However, these need to be complemented with other classification systems as the services are expanded more into health-related social services. For the general overview analysis, the following questions can be used to assign a criterion score:

- Are all categories of diseases, operations, injuries, and so forth handled by the organization? There is no need for your organization to provide all the services directly. You may have an arrangement to address all the categories through some contractual or affiliation relationships, as part of your VIHS.
- Have you identified duplications or conflicts that need to be addressed?

In the comments section, you may want to note the gaps of topics not addressed, duplications, conflicts, and issues. For an existing or proposed system, it may be inappropriately duplicative for each partner to offer the same services. The key is for the system in some manner to address all the health and healthcare needs of the communities served.

Foci

Fourteen different foci are illustrated in Figure 4.3: promotion, protection, prevention, detection, diagnosis and assessment, treatment, habilitation and rehabilitation, maintenance, hospice, support, advocacy, education, research, and enabling. Very few organizations address the full range of foci. Hospitals, for example, have traditionally focused almost exclusively on diagnosis and treatment. The score for each criterion can be based on questions like the following:

- Does the organization have any recognized, budgeted services that address this focus?
- Has the organization achieved any measured and recognized success for these efforts?
- Are there inappropriate duplications or conflicts?

If an organization does not have any regular services or programs to focus on health promotion, it may score zero on this criterion. If the organization has an active hospice service that offers services to the families of all patients who are dying and provides services to a large percentage of those patients and families, then the organization may receive the highest score, 4, on this focus. In the comments section, you may want to note specific areas for follow-up.

Settings

Eleven categories of settings are illustrated in Figure 4.3: areawide, community, houses of worship, schools, work, mobile, home, ambulatory, partial-day care, inpatient, and freestanding support. The score for each criterion can be based on questions like the following:

- Does the organization offer services in this setting?
- Has the organization achieved any measured and recognized success for work in these settings?
- Are there inappropriate duplications or conflicts?

One of the questions about areawide services is whether advertising constitutes services provided on an areawide basis. If the organization is simply running advertisements to attract people to its facilities, then it would score low on the areawide criterion, although the ads may indirectly raise awareness. If the aim is prevention or health promotion, then the score for the areawide focus would be higher. An example would be a healthcare organization that is jointly working with schools and other community organizations to reduce injuries resulting from drinking and driving. Again, the comments section may be used to identify topics and issues for follow-up.

Core/Key Processes

Nine core/key processes are illustrated in Figure 4.3: leadership and governance, strategic planning, human resources management and development, process and quality improvement, information planning and management, continuum of care, client/patient rights and satisfaction, prevention and education, and managing the environment. Your organization may want to add one or more core/key processes to this list, based on your unique situation. Since the expectations of external organizations are organized into core/key processes, the scoring of these processes is addressed in Chapter Six. These processes could be scored in either place but should not be scored in both, since they reflect the same measures.

Resources

Five broad categories of resources are illustrated in Figure 4.3: human resources, physical resources, financial resources, organizations, and information. There are specific resources, or criteria, listed on each. The score for each criterion can be based on questions like the following:

- Does the organization provide this resource?
- What is the relative value of this resource? For example, can another resource provide the required service more cost-effectively?

Since different resources can provide the same or similar services, a key question regarding the analysis is the appropriateness. However, this judgment can be made only at the more detailed levels of analysis. The first step is to identify the existence and general capabilities of resources.

Steps for Using the Analysis Tool

Now that you understand the approach to conducting the analysis and the use of the tool, your organization should tailor the process and tool to meet your requirements:

1. Decide on the purpose of the analysis.

   Are you analyzing a single organization?

   Are you analyzing an existing health system?

Are you analyzing potential partners in a health system?

Are you comparing against competitors, or system partners, or both?

2. Determine the level of detail for your current assessment: overview, intermediate, or detailed.

3. Review the criteria used in Figure 4.3 for each dimension of the taxonomy. Do you want to add, delete, or amend any? Remember the caution here: using a higher level of detail is fine, but deleting a component of the taxonomy may cause you to fail to meet the intended completeness of the analysis using the taxonomy.

4. Revise the analysis tool to meet your requirements.

5. Select the people to participate in the analyses.

6. Review the analysis tool, the operational definitions of terms, and the scoring method with all the participants, to reduce variation among different people using the analysis tool.

Conducting More Detailed Analyses

The overview analysis typically is based primarily on perceptions and readily available data, with minimal effort spent collecting data. The risk, of course, is that the perceptions may be incorrect, so more detailed and quantitatively based analyses may be desired concerning selected topics, especially those portions of the taxonomy for which scores from the overview analyses show the greatest need for improvement or those that relate to tentative plans of your organization. The same analysis tool, Figure 4.3, can be used. The difference is that now the scores entered for each criterion are based on more specific and quantitative data. As an example related to settings, locations of different services for your organization and the comparison organizations can be plotted on maps or assigned to geographic grids. This information would then be used to make more detailed assessments of the appropriateness of duplicate services and facilities.

Based on the overview and intermediate-level analyses, certain aspects of the scope of services may be selected for detailed analyses.

Actions Based on Analyses

The scope-of-services analyses using Figures 4.1, 4.2, and 4.3 will lead to a number of actions. The following are examples of the types of actions and additional analyses that your organization may undertake:

- Identify opportunities. By the time you have finished the analyses of scope of services using the taxonomy, you will almost certainly identify some opportunities for new services, changed foci, additional settings, or different resource mixes.
- Identify duplications. You need to identify duplications within your system, or among your organization and potential partners. Some of these duplications will be unnecessary and costly; others will be appropriate.
- Determine the appropriateness of the current distribution of foci, settings, resources, and so forth. Appropriateness is difficult to evaluate, because different people have different perceptions of appropriateness. Basing this determination on the healthcare, health, and social services requirements of the populations served is an effective way to answer the appropriateness question. You can begin with demographic and health status information about the populations in the communities served and then estimate use rates of different services per 1,000 population (for example, acute inpatient days or service-specific outpatient visits per 1,000 population, number of family counseling visits per 1,000 population, number of school visits per 1,000 students, or time to reach a primary care provider). Care must be used in expanding beyond historical use rates and exploring alternative configurations of services from the taxonomy. If you rely solely on historical use rates, you will replicate existing practices. You should look for alternative models to meet population needs that improve quality, outcomes, and cost-effectiveness, not just duplicate past patterns of care. Identifying best practices as benchmarks is an important step. This will allow you to determine criteria and thresholds for the settings, numbers, and locations of services. Another useful approach is to estimate use ratios, and the resulting resource requirements, based on multiple, alternative configurations.
- Identify issues and conflicts for resolution.
- Assess compatibility of potential partners.
- Identify and prioritize areas for more detailed analyses.
- Identify approaches to fill gaps.
- Identify approaches to eliminate inappropriate duplication.
- Develop collaborations to maximize quality and cost-effectiveness.

The results of the analyses related to scope of services, using the taxonomy, are among the key inputs to assessing or developing a VIHS. The expectations criteria of external organizations are discussed in Chapters Five and Six, and internal organizational climate is discussed in Chapters Seven, Eight, and Nine. Consolidation of the three types of analyses is discussed in Chapter Ten.

CHAPTER FIVE

ANALYZING EXTERNAL EXPECTATIONS

Strategic planning for integration should start with the identification of your organization's strengths and weaknesses, as well as your system's collective strengths and weaknesses. Strategic planning for integration should also include evaluating the organizations you are planning to integrate with in order to assess their strengths and weaknesses, as well as evaluating competitors within the environment. The entire system would benefit from an assessment of strengths and weaknesses in its key processes. External expectations and self-assessment are a good place to start (see Figure 5.1). This chapter analyzes many of the external expectations for health and social services organizations and demonstrate how they can assist leaders of merging healthcare systems in terms of strategically planning for integration. The use and evaluation of external expectations is easily adapted to the analysis tool introduced in Chapters Two and Four.

Whether a single healthcare or social services organization or an integrated system, one of the common elements is the role of external review bodies in evaluation. Review, licensure, accreditation, or recognition by a representative body or bodies has been a necessary but often painful and frustrating requirement for many different organizations. It is one of the common experiences that people from different organizations can immediately agree on. Unfortunately, a common feeling among people within organizations is that many times review bodies have failed to prove their ultimate worth. Through punitive surveys, poorly trained surveyors, and poor communication of vision and direction, it has been difficult to

FIGURE 5.1. EXTERNAL EXPECTATIONS.

A pie chart with six segments: External Expectations, Internal Organizational Climate, Financial Analyses, Legal Analyses, Taxonomy of Services.

separate the value from the process. Reviewers, inspectors, or surveyors may have little knowledge of current industry pressures or may possess a strong bias toward their way. There have been complaints concerning the inflexibility of reviews in an industry that has to become more flexible to survive.

Our collective experience has been that most organizations go through the motions of the review process simply to get it over with, without ever stepping back and analyzing how to improve. External review is often seen as an imposed industry standard, forced on an organization and not possessing any inherent value. For example, funding is often tied into reviews. Medicare, Medicaid, Blue Cross, and other payers will not reimburse a hospital unless it is accredited, by either the Joint Commission on Accreditation of Healthcare Organizations (JCAHO) or the Health Care Financing Administration (HCFA). For many organizations, the goal becomes merely to pass. In this chapter, we are going to look beyond the multiple review processes, into the role and use of external expectations in planning for and executing integration.

Our approach is to set aside the whole external review process as it is conducted by these organizations and truly look at and analyze the requirements. We examine how to use them to approach and plan for expectations. This strategy then provides twice the value: in using and incorporating external review expectations, a single organization or an entire system will use common criteria for evaluation and continue to meet the expectations of formal review organizations. Valuable information about actions and evaluation is contained within these standards.

It is common for people to react negatively when someone mentions the value of something like HCFA surveys, state and local department of public health (DPH) visits, or the JCAHO. Be reminded of a simple fact: many of the review bodies discussed in this chapter were created, financed, and staffed by social services and healthcare insiders, such as social workers, physicians, therapists, nurses, and many others from our very organizations. Institutions such as the Council on Accreditation of Services for Families and Children (COA), the Commission on Accreditation of Rehabilitation Facilities (CARF), and the National Committee for Quality Assurance (NCQA) do not create their requirements in a vacuum. They convene groups and panels so that social services and healthcare industry members, customers, and practitioners can decide what is important and what needs to be evaluated. The COA, which accredits public and private child welfare programs, establishes criteria around the assessment of its clients because client assessment is considered a critical function with tremendous impact on the outcome of treatment. CARF also establishes criteria pertinent to the assessment of clients for the same reason. The focus of these review, licensure, accreditation, or recognition organizations may be broad or narrow, but the criteria for review and evaluation represent the standards expected by health and social services organizations in the industry. The outcome reflects a national consensus on significant organizational processes in different social service and healthcare organizations in the industry.

Many types of external review bodies have spent a great deal of time, effort, and money in the past several years to create meaningful review processes and evaluation standards. The one thing that has not happened is the demise of the mind-set that this is "just another thing to get through before we get down to business." Actually, the organizations that get the most out of external reviews are those that recognize that the criteria are a reflection of the right things to do to meet customer needs and remain viable. Doing well during the review is secondary and is simply another indicator of overall success.

One undeniable fact is clear: many of these established review bodies carry with them credibility. The JCAHO, the COA, and the NCQA are three of the most well-known and influential of the accrediting bodies in health and social service organizations, and for those outside the field, they represent the official determiners of quality. Given the lack of outcomes data or other quality-related information, how else can organizations not involved in healthcare make judgments about quality? Few payers are interested in reimbursing a nonaccredited facility. Knowing this, some well-known healthcare organizations offer incentives to their leaders for achieving high scores during review. One can only assume there might be a distinct advantage to this tactic. We offer the following discussion and tool, which is discussed in detail in Chapter Six, as another strategy and a good place to start. It fully exploits the value and utility of very often expensive external review processes.

There are many common key processes identified within the various requirements. The common ones are leadership and governance, strategic planning, human resources management and development, process and quality improvement, information planning and management, continuum of care, client/patient rights and satisfaction, prevention and education, and managing the environment: building, people, and equipment safety. It is hard to disagree that these are important processes that any organization or system of organizations should focus on when planning for a successful business. Sound knowledge of organizational processes such as the ones listed above provides a basis for evaluation, discussion, and planning of an individual organization, and certainly a potential integrated system of social and healthcare organizations.

For years, people within these social and healthcare organizations have been accountable for ensuring that the organizations meet all the requirements of the review bodies. This is especially true in some healthcare organizations, such as hospitals, where whole departments or job functions may have been formed to ensure competency in the requirements. Tremendous resources have been focused on passing inspection. Traditionally, meeting external requirements and undergoing inspections or surveys for certifications have been annoying, time and resource consuming, and plagued by nervousness and doubt. We believe, however, that the vast majority of social and healthcare people would admit (albeit grudgingly) that these external review bodies have helped make them a better organization. We are suggesting that your organization or system move past the episodic reviews.

How many times have the requirements been assessed and analyzed for internal use? Can you argue that these standards and requirements reflect the very same processes an organization must pay attention to in order to maintain a competitive, high-quality, efficient organization? Objective assessment, based on key functions and processes important to your customers, is a major strategic step in moving toward organizational improvement and integration. Evaluating your organization against how well it performs the processes identified within the external criteria can reveal where organizational strengths and weakness lie. The value of an objective assessment is that it gives the organization or system an initial idea of priorities for improvement. Seldom, however, have we seen any linkage among the external expectations that need to be met and strategic planning.

As organizations consider integration and participation, this type of an assessment is a good place to start in order to document your performance for a potential partnership. There are people within each organization, as well as external experts, who can provide this type of evaluation. Since the future of VIHSs en-

compasses a much broader set of services, it may also help to understand organizations you may not be a familiar with, such as social services agencies. Formal survey or informal survey, the plan is to evaluate the most important function
an organization performs in an objective manner.

Business leaders, payers, and consumers typically have little information by
which to compare healthcare organizations. This self-assessment also can serve as
documentation of a focus on performance to your customers. External criteria
and review have a history of being a proxy for quality, particularly for payers and
organizations not in the industry. Through the definition of key processes within
these expectations, the external review organizations have attempted to define
quality. Many managed care organizations require accreditation from one of
the established external review bodies such as the NCQA or the JCAHO. In reality, we are really suggesting two compatible ideas. First, external accreditation
is an important and necessary part of any social services or healthcare organization. Most organizations need the stamp of approval from most of the appropriate bodies. Second, beyond the stamp, there is value in moving past the mechanics
of achieving the certificate and using those standards as guides for truly focusing
your efforts toward quality and efficiency. Identifying opportunities for improvement, as well as highlighting marketable strengths, are important reasons for using
external expectations.

Customer Satisfaction

Gaucher has said that "one of the keys to achieving customer satisfaction is communication" (Berman, 1995, p. 253). Customers may not verbalize all requirements, but when asked, they will verbalize their perceived requirements. Kano
(1984, p. 6) described three different measures of quality, as illustrated in Figure 5.2: apparent quality measures, exciting quality measures, and expected quality measures.

Apparent quality measures, like waiting time, are readily perceived, and customers
are often sensitive to them. As the measure is better fulfilled, customer satisfaction
steadily increases. These measures are normally mentioned when customers are
asked about their expectations. On the other hand, the other two types of quality
measures will seldom be mentioned when customers are surveyed. *Exciting quality*, such as a staff person's stopping work to assist a lost visitor, is seldom mentioned
during surveys or interviews because the customer may not have experienced this
type of service before. Yet this type of quality produces delighted customers. *Expected quality* is also seldom mentioned during inspections because customers expect this level of service.

FIGURE 5.2. KANO MODEL OF QUALITY MEASURES.

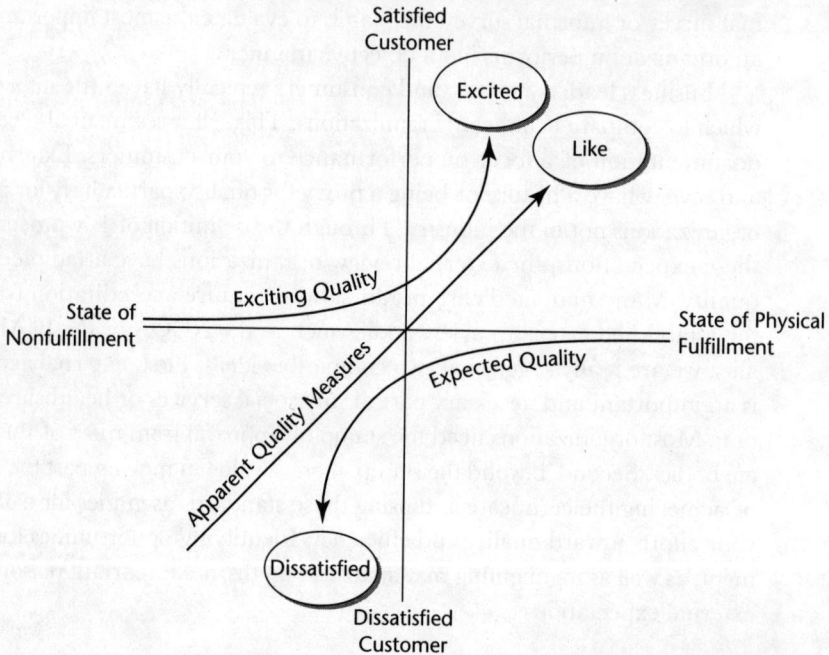

The point is that you must consider not just expectations expressed by patients or clients but also expected and unexpected quality items not expressed by customers. For example, patients will seldom mention the avoidance of a hospital-acquired infection as a measure of quality because patients expect that they will not acquire an infection while in the hospital. Health and social services professionals normally establish professional standards related to expected quality—a safe environment, monitoring housekeeping and infection control practices, and the credentialing of caregivers—which patients are unaware of.

Healthcare and social services organizations have traditionally not held themselves to meeting minimum standards of quality. Rather, they have invested time and money to demonstrate a commitment to a higher evaluation of our industry. We have created a complex matrix of expectations and outside evaluators, and actively participated in this system by crafting an external review process focusing on much higher standards. These bodies have identified various functions and processes that are important to remaining the highest-quality healthcare system in the world.

It does not matter at what level the evaluation begins: single organizations, networks, purchasers, or payers. These evaluations will provide a forum for discussions around the processes and goals set out by various review bodies and some interesting insight and guidance toward a productive and meaningful discussion.

The External Review Bodies

Virtually all social services and healthcare organizations have some external expectations that they must meet. In fact, large systems may be reviewed and inspected by forty or more external bodies. The *Lexikon*, developed and published by the JCAHO (Joint Commission on Accreditation of Healthcare Organizations, 1994), contains the most comprehensive list available of the review, licensing, accreditation, and award organizations with an impact on social services and healthcare organizations. Most of the organizations listed were drawn from the 1993 edition of *Encyclopedia of Associations*.

We briefly discuss several of the external review and award bodies in healthcare and social services organizations. We have chosen a representative sample of different types of review bodies that focus on various areas (for example, managed care plans, rehabilitative service organizations, child welfare and social programs, ambulatory organizations, home care services, hospitals, and networks or systems of care). We include no governmental regulatory agencies since their expectations are encompassed within the expectations from the organizations mentioned. We also include the Malcolm Baldrige National Quality Award for healthcare organizations; although it is not an external review body, it is widely considered the preeminent award for quality.

Malcolm Baldrige National Quality Award for Healthcare Organizations

The Malcolm Baldrige National Quality Award of 1987, signed into law by President Reagan, was designed to promote awareness of the importance of quality improvement to the national economy. The award promotes awareness of quality excellence, recognizes quality achievement of U.S. companies, and publicizes successful quality strategies.

Initially the award precluded not-for-profit organizations from applying, but on February 3, 1995, an adaptation of the Baldrige criteria, which is to run parallel with the Baldrige Award for businesses, was released for piloting in the healthcare industry. The category is limited to healthcare organizations that primarily engage in providing healthcare services to people. The areas of focus for the award are leadership, information and analysis, strategic planning, human resources development and management, process management, business results, and customer focus and satisfaction. Recently, the Baldrige has changed its focus from quality at the expense of everything to quality plus financial results.

The award is still being tested and has not yet been activated. Many healthcare organizations have requested the criteria in order to incorporate an established set of guiding criteria into their changing organizations. The Baldrige Award and its state and local offshoots have been key in the effort to strengthen U.S. competitiveness and to galvanize the U.S. quality efforts (Malcolm Baldrige National Quality Award Office, 1995).

Accreditation Association for Ambulatory Health Care

The Accreditation Association for Ambulatory Health Care (AAAHC), once a review arm for the JCAHO, became an independent accrediting body in 1979. The board comprises several of the ambulatory care specialties. The mission of AAAHC is to assist ambulatory healthcare organizations in improving the quality of care provided to patients. The association operates voluntary, peer-based accreditation programs and consulting services. It establishes standards, measures performance, and acts as both a consultant and educator to participating organizations (Gonen and Probyn, 1996).

Community Health Accreditation Program

The Community Health Accreditation Program (CHAP) is a small but growing body, focusing on accreditation of home- and community-based programs. Established as an arm of the National League for Nursing, CHAP received deemed status from Medicare for home care services prior to the JCAHO's home care program. Deemed status is the designation conferred by HCFA on a healthcare provider when that provider is judged to be in compliance with Medicare conditions of participation because it has been accredited by a voluntary organization whose standards and survey processes have been determined by HCFA to be equivalent to those of the Medicare program. This means that home care programs wishing to receive Medicare reimbursement could use CHAP accreditation in lieu of Medicare certification (Gonen and Probyn, 1996).

Council on Accreditation of Services for Families and Children

The COA is an independent not-for-profit organization established in 1977. It accredits approximately one thousand behavioral healthcare programs and three thousand social services programs delivered by nearly seven hundred providers in the United States and Canada. The COA plans to publish standards for networks made up predominantly of child and family services agencies, as well as standards for Canadian providers.

The COA has standards for over sixty behavioral healthcare and social services, including outpatient mental health and substance abuse services, services for persons with disabilities, services for victims of domestic violence, foster care and day care services for children, case management, day care services for the elderly, and residential treatment for youths. The mission of the COA is to strengthen and promote the quality of social and mental health services that support and improve the lives of families and children and the well-being of society (Council on Accreditation of Services for Families and Children, 1996).

Commission on Accreditation of Rehabilitation Facilities

CARF is the not-for-profit, national accrediting body for organizations providing rehabilitative services. CARF was established in 1966 through the support of many national organizations representing a broad range of rehabilitative services. Since then, more than twenty-three sponsoring member organizations and more than eighteen associate member organizations have joined together in support of CARF's goal to become the preeminent standards-setting and accrediting body for organizations delivering services to people with disabilities and those in need of rehabilitation (Gonen and Probyn, 1996; Commission on Accreditation of Rehabilitation Facilities, 1996).

Joint Commission on Accreditation of Healthcare Organizations

The JCAHO is one of the oldest and best-established external accrediting organizations. Established by several physician groups in 1951, the JCAHO began as an accrediting body for hospitals. In 1965, the organization received deemed status for hospitals. This meant that hospitals wishing to receive Medicare reimbursement could use JCAHO accreditation in lieu of Medicare certification. Most hospitals chose to be accredited by the JCAHO. Later, the organization branched out into other healthcare avenues and now accredits more than fourteen thousand healthcare organizations, including hospitals, home care providers, behavioral healthcare organizations, long-term care facilities, ambulatory providers, laboratories, and, recently, healthcare networks.

The JCAHO was interested in accrediting managed care but dropped the program in 1989 due to lack of interest in the market. In 1992, the organization resurrected the program, and it drafted the initial standards in 1994. The JCAHO standards for managed care apply to healthcare networks, defined by the JCAHO as "entities that provide, or provide for, integrated health services to a defined population of individuals." The program includes tracks for HMOs, preferred provider organizations (PPOs), independent practice associations (IPAs), full-service networks, and specialty networks.

In 1987, the JCAHO launched its Agenda for Change, which refocused the standards and survey process toward a functional, integrated approach. The standards no longer reflect a focus on departments and services but rather on the processes an organization performs that affect patients as they move through the organization. The functions and standards address clinical and administrative functions. In 1994, performance reports for individual accredited organizations were made publicly available (Joint Commission on Accreditation of Healthcare Organizations, 1996).

National Committee for Quality Assurance

The NCQA was formed in 1979 as a joint effort between Group Health Association of America and the American Managed Care and Review Association. The purpose was to promote quality assurance, standards, and performance measures. In 1991, the NCQA, now independent, reviewed and accredited more than 300 of the nation's more than 574 HMOs. The board of directors includes consumers, labor, purchasers, and health plan representatives.

Health plans apply for accreditation if they provide comprehensive services through a defined delivery system to a specific population and have been operating for at least eighteen months. The survey is conducted by a healthcare team consisting of physicians, nurses, and administrators. NCQA accreditation has emerged as a nationally recognized evaluation that purchasers and consumers use to assess managed care plans. The NCQA has become better known of late, due mainly to the endorsement the organization received from the American Association of Retired Persons (AARP). HMOs and managed care plans that accept Medicare populations will be required to have NCQA accreditation (National Committee for Quality Assurance, 1994; Gonen and Probyn, 1996).

Other Organizations

The organizations we have identified represent only a fraction of the total population of external review bodies. Organizations such as OSHA and HCFA are included among the external review bodies. Most of the expectations from these broader regulatory agencies have been carefully and purposefully incorporated into many of the standards of the accrediting bodies mentioned and those that were not mentioned. Remember to include local and state idiosyncrasies, or check to see if they are incorporated into some of the national external review expectations.

There is a growing recognition by purchasers, payers, and providers of the need to incorporate alternative approaches to health. This industry represents a huge portion of the dollars currently spent on healthcare. The Alliance for Alternatives in Health Care is a national organization, established in 1983. Its membership is composed of over a thousand individuals seeking recognition from the public and healthcare personnel of an alternative approach to medicine. The focus includes, but is not limited to, chiropractic, holistic, homeopathic, acupuncture, and natural treatments. Strategies for partnering with and offering alternative health approaches may prove to be very beneficial in the near future.

We mention another type of organization that should also be incorporated into the assessment process, although it may not offer external review, promulgate standards, standards measures, or guidelines. These include organizations such as Agency for Health Care Policy and Research and the Institute for Quality Improvement, as well as systems that have developed clinical indicator systems such as Cleveland Business Coalition, Maryland Indicator Project, and the JCAHO. There are several existing and developing groups that can assist with assessment efforts.

Crosswalk Development

As a result of our work, we have identified the huge overlap among external expectations, and it did not take long for us to decide that all of these expectations could be compared, using a series of *crosswalks*, or grids that graphically show overlapping expectations. So we set to developing both high-level and detailed crosswalks between the various external review organizations. Although there are multiple review, licensure, accrediting, and recognition bodies to contend with, a relatively small and succinct set of common expectations overlaps in all of them. These crosswalks have helped us demonstrate that allocating resources for each review organization is a waste of time and money.

We advise that your organization and partners begin to incorporate all of the various external expectations each individual organization is subjected to into an integrated, process-oriented approach. These crosswalks helped identify nine dimensions of performance that organizations assessing themselves, potential partners, the system, or competitors will want to evaluate.

We have reviewed and consolidated dozens of external expectations and distilled them into nine dimensions, with multiple criteria under each dimension. Appendix A contains a sample of the high-level crosswalks we initially constructed, which demonstrate the overlap in expectations. These crosswalks show only the categories where overlap occurs. For example, one of the categories, or dimensions, examined in the Baldrige Award process is "Focus on Patient/Stakeholder

Satisfaction." This category, and the criteria within it, focus on the same aspects as the NCQA standards under "Quality Management and Improvement" and "Member Rights and Responsibilities." The high-level crosswalk assigns categories or dimensions to corresponding categories or dimensions. The detailed crosswalks, which are much larger, include standards and expectations crosswalked within and among each dimension.

We strongly recommend that external review organizations begin to coordinate among themselves in order to standardize review criteria to the extent possible. This action will result in reduced confusion, time requirements, and costs to the entire health and social services system. If we are moving toward a VIHS, we have to focus on a virtual set of expectations for this system and get rid of duplicative, burdensome, and fragmented expectations, which add to an already overburdened and redundant approach. This move may eventually result in a consolidation and a dismantling of many of the external review bodies, but how is this different from what their customers in healthcare are experiencing right now? It should not be surprising since the external review organizations are actually a reflection of the organizations they serve.

Virtual Set of Expectations

Reviewing many of the external expectations offers social services and healthcare delivery systems a common set of evaluation tools when considering whom to integrate with and how to approach the initial plan for integration. These can be used as guidelines when creating new services and improving or changing existing services. The evaluations can serve as a guide to aid in planning and evaluating current or potential partners in your emerging delivery system. Just as the taxonomy presented in Chapter Three helped create the blueprints for the system, the discussion around important functions and processes can serve as building material for your system.

We briefly describe the typical dimensions, or functions, to clarify the interpretation. The relative priority of these different dimensions will vary among organizations. Your organization may wish to amend, add, or subtract from the list. However, you need to ensure that the functions that your organization is subject to are incorporated into the functions listed or add them to the list.

The external expectations a health or social services organization must, or chooses to, meet can act as building blocks of evaluation and integration planning. It is possible to identify common, important, organizational characteristics, clinical and nonclinical, through external standards. Each external body may place greater emphasis on a particular requirement or focus for the organization. For example, the Baldrige requirements focus more heavily than others on the information management aspects of an organization. Many accrediting bodies place greater emphasis on clinical and administrative functions of healthcare organizations. HCFA, DPHs, OSHA, and others tend to focus on safety and environmental issues. Each external body focuses more heavily on certain aspects, but there is agreement on the whole about which are important processes and how they should be addressed.

The following nine dimensions, identified and incorporated into the analysis tool, are important to your system to use in evaluations:

1. Leadership and governance
2. Strategic planning
3. Human resources management and development
4. Process and quality improvement
5. Information planning and management
6. Continuum of care
7. Client/patient rights and satisfaction
8. Prevention and education
9. Managing the environment: buildings, people, and process safety

These nine dimensions emerged from our analysis and consolidation of multiple review, licensure, accreditation, and recognition expectations. The categories within the separate review bodies, although titled differently, are captured under one or more of these dimensions. Criteria and appropriate questions to ask under each of these dimensions are discussed in Chapter Six, where we will apply the dimensions along with criteria for each to the analysis tool.

Leadership and Governance

Leadership and governance in organizations are leading indicators of organizational success. Effective steps in the leadership process are identified throughout the standards. The role of governance is important because the governing body is assigned the role of link between the population served and the health and social services organizations. *Common direction* is a term used to refer to the existence and alignment of the mission, vision, values, critical success factors, goals, key processes, key indicators, and reinforcement through reward and recognition

Leaders are asked to create and communicate a compelling vision of the organization. Values are established, guiding how these organizations will function together. Clear expectations and true commitment to quality for the organization are also identified throughout these expectations. The standards may focus on how the leadership integrates mission, vision, values, and quality into the entire management structure of the organization. The governance and the organizational leaders are reminded to consider the community as well in their planning efforts.

Strategic Planning

The process of strategic planning for an organization is important to identifying and meeting the expectations of the organization's primary customers, especially in today's environment of rapid change. These expectations describe the components that leadership may consider in the strategic planning process. Strategic planning provides direction and milestones toward organizational mission and vision fulfillment.

Essentially, you must answer three questions: "Who are we?" "Where are we going?" and "How do we get there?" It is important that there be alignment of these questions from the corporate to departmental to personal levels. It is also important that behaviors be consistent with the written statements, to achieve success. Gaucher and Coffey (1997) provide a more extensive discussion of organizational alignment and how it is measured.

Strategic planning encompasses service design and redesign and resource allocation. Any system requires a reliable, solid financial planning process in order to survive. Some of these components consist of incorporating patient requirements and expectations, clinical personnel expectations and requirements, and changes in the environment and competitors.

Strategic planning and common direction are particularly critical during the transition from fee-for-service, through fixed-price-per-episode (like DRGs), to fixed, capitated reimbursement pmpm. During this transition, the financial incentives, and consequently the planned services, change radically. Under a fee-for-service system, whether fully reimbursed or discounted, the incentive for healthcare providers is to provide more services and gain market share of the services provided. There is little attention paid to population health and social and environmental conditions, because these are traditionally not reimbursed. Under a fixed-price-per-episode system, the provider incentives are to minimize the amount of costs expended to provide care for each episode but to expand the number and market share of those episodes.

Under a capitated system, the incentives for providers are to gain market share of insured lives, particularly those that are low utilizers of expensive healthcare services, and to provide far fewer expensive healthcare services. The incentives for payers and employers under the fee-for-service and fixed-price-per-episode systems conflict with those of the providers. These transitions create completely different strategies. Hence, the strategic planning to establish a common direction and clearly communicated goals are critical.

Strategic planning criteria focus on assessing organizational or system capabilities and also consider future risks of many kinds. Strategic planning requires actions to carry it out successfully, as well as continual evaluation and improvements to the implementation.

Human Resources Management and Development

Review bodies identify the dimension of human resources as extremely important in any organization or system. This is true in any type of business organization and is particularly important in areas that require complex, highly skilled, and technical resources who must also work with the patients and clients. The goal is to provide the organization with the right number and kind of people. The delivery of healthcare services requires highly skilled, competent people. It requires tracking and credentialing of practitioners and a focus on continually improving the competence of the entire staff.

How an organization maintains a productive work environment is also a focus. Employee satisfaction is solicited and incorporated into the planning and delivery of services within the system. Organizational leaders are asked to consider the environment in which the employees work. Evaluation of the alignment of employee development with organizational goals is a priority. Education and training are planned for, delivered, and evaluated in a manner that contributes to both employee and organizational growth and well-being.

Process and Quality Improvement

Many external bodies have incorporated the fundamentals of Total Quality Management or Quality Improvement into their expectations. These expectations tend to provide a framework for the other dimensions and criteria within the dimensions, so that quality is incorporated into everything the system does. The foci of the quality expectations are on the external environment, internal processes, and the need to incorporate monitoring and evaluation into these processes. Two subsets in this dimension are important:

- Operations and service design and improvement. The processes of design and management for operations and services are key to ensuring high quality and cost-effectiveness. The better that processes are designed to meet customers' expectations and desires, the greater the customer satisfaction that will be achieved. Similarly, the better the operational processes to provide services and products, the better the quality and cost-effectiveness.

- Measurement and monitoring. An old adage is relevant here: "You can't manage it if it isn't measured." Performance measurement and monitoring processes are vital within any integrated delivery system. Performance management provides feedback and guidance for system-wide, institutional, services, or individual improvement and must be addressed on multiple levels simultaneously.

A major challenge to any VIHS is to expand the integration and time window of measurements. Traditionally, for example, acute care hospitals and physicians measured quality and cost only while a patient was an inpatient in the hospital. More recently, the measurements have been expanded to cover a short period before and after an admission or procedure. With the expansion to address health and cost pmpm, overall costs, functionality improvements, and perceptions should be measured across providers, geographic settings, and time.

The focus is extended to the use of data to support organizational quality in performance and as a business entity. Leaders are asked to identify and understand how information and data from all over the system are integrated and analyzed, then incorporated into organizational planning and decision making.

Information Planning and Management

The most intimidating, complex, and possibly most important dimension that health and social services must contend with now and in the future is successful management of information. Every process we perform—in fact, everything we do—is based on data and information. The processes of planning for and managing information and communication are extremely important, particularly as more foci, settings, and resources are involved. The obvious place to begin to merge information systems and processes is at the level of the patient. The continuity and coordination of care is but one of the important communications necessary within an integrated system. For true care coordination, information is needed at four distinct but interrelated levels.

Individual-Level Information. Data on individuals are necessary to manage their care, services, and health. During the formation of an integrated health system, one common problem is developing coordination among existing records of different healthcare, health, and social services providers. Medical record numbers may be different wherever the patient enters the system. Case numbers for clients who enter into healthcare organizations from or in addition to social services organizations may not be communicated and incorporated, information stored is different, the operational definitions of information are different, and the information systems are different and incompatible. To provide continuity of care, information about individual patients or clients must be available at different care sites, when and where care is rendered.

Aggregate-Level Information. Aggregated data are necessary to make decisions about how best to manage healthcare, health, and social services. The focus shifts to the clinical or care component. Aggregate data from clinical trials, for example, are required to determine the best care protocols. Information must be available from social services and healthcare about care processes. Information about the services your organization provides must be standardized, because it is through standardization that improvement and quality occur.

This concept of standardization appears to be the antithesis of much of social and behavioral health approaches because the focus within these services has traditionally been on individualization for clients. It is just this type of approach that has frightened, and in many cases precluded, payers from incorporating these types of services and treatments into a reimbursement scheme. Social services agencies must begin to develop and promote standardized criteria, for example, the cycle time effect of foster home placement on children within the system. Is there a real cost benefit to screening in schools for children of abuse? Does this possibly reduce the cost of emergency room visits, not to mention the emotional impact on the child? Developing real-time, interactive database management capacity will speed network and systemwide coordination of services.

System-Level Data. The third type of data are those directly related to the system, their components, and their providers; these data focus on the business portion of the system. Separate from care information, these data provide vital and necessary business information for organizations and systems. A set of information needs to be available on billing, cycle times, material services, and facilities.

Community-Level Data. Knowing the community served, including the population and other demographics of the community, will be of immense value to systems that seek to predict accurately the needs of the population they serve. Utilization of different community data will allow successful, long-term strategic planning and initiatives that have a much longer time interval. Systems will require advanced knowledge about the population demographics, public health measures, employers within the community, and other potential resources.

Health organizations are highly data dependent. Because of this, most of the external review bodies have expended a great deal of time and other resources into identifying steps in the process of planning for and developing successful information management systems. In developing a system, it is useful to begin by identifying the information needs. This process of a needs assessment encompasses both internal data needs and external expectations. The goal is to gather the data using a variety of sources, such as employees, patients, and suppliers. The information must be accessible and utilized by practitioners, administrators, and others to improve the delivery of care.

Information should be useful to the governing body and other support services so that they can adequately plan for and support the clinical practitioners. One of the biggest obstacles in integration is the planning and execution of a comprehensive, seamless, useful information system. Most organizations have some type of system, but attempts to combine systems result in failure. The single most important step in integrating information systems that are of high quality, comprehensive, and accessible is the development of a single plan based on a needs assessment. The development and use of common operational definitions, or tables to convert operational definitions, is crucial in the integration of information systems.

Confidentiality and security are important issues addressed throughout external expectations. These are increasingly more difficult and important as services and information are geographically distributed. There is also the consideration of making sure the literature required for decision making is available. Finally, there is a focus on pulling all the information together and using it in patient and other client treatment at any time and any place they may enter the health or social system.

Communication is considered an integral part of managing information just as it was under the leadership dimension. It is a vitally important process—the real key to success. For example, each of the following components is important to communicate:

- Mission, vision, values, goals, and common direction of the organization, particularly important during the transition from fee for service to capitated reimbursement.
- Individuals' roles and responsibilities toward the organization's common direction.
- Case management guidelines, treatment protocols, care maps, and critical pathways, to establish common understanding, expectations, and standards of care. As we focus more on maintenance and improvement of health, protocols should be developed to assist people in managing their own chronic health conditions, such as diabetes and hypertension, or identifying early

intervention strategies for high-risk populations from both a health and social service perspective.

- Educational materials that promote healthy behaviors and cessation of risky behaviors.
- Educational materials that ensure a common understanding and competence among the providers.

Continuum of Care

Assessment and intake, treatment planning, and the continuation of care into another setting or focus of the individuals are addressed within this dimension. Planning, implementing, and monitoring of new and existing services into social services and healthcare organizations are also the foci of many of the external bodies. The goal is a seamless and coordinated continuum of care. The standards examine the key aspects involved in process design and management. The process includes both planning for and designing patient care services, utilizing patient care support services and community resources, and integrating administrative processes into the design and delivery of services. Many identify the importance of planning and managing supplier or contract services. These expectations call for the involvement of all staff (clinical and nonclinical), customers, patients, suppliers, and community in all levels of operations. The organizations meeting these expectations are asked to describe how patient and other client care services are being managed and monitored. The goal is to ensure that quality, effectiveness, and efficiency are built into processes and continually assessed and improved. Two components need to be addressed.

Access to Services. The processes to ensure access to services occur at two levels. First, each organization must address a population's eligibility for services, including those with insurance, covered lives, and the uninsured. This encompasses the mechanisms to communicate the availability of services and ensure that people can receive services. The access question is much broader than simply taking care of people when they arrive. Availability of transportation, hours of service, and cultural barriers are significant issues related to access. Second, access to appropriate services within the overall community should be addressed. This area has seldom been addressed because each organization establishes its own access processes, based on its own incentives, with little or no coordination among organizations. System formation may hasten resolution of many access issues within the communities served if planning is based on needs of the entire population.

Intake, Admission, Registration, and Scheduling. Closely related to the access question are the intake, admission, registration, and scheduling processes. The term *admission* is generally used regarding inpatients, and the term *registration* is generally used regarding other types of services. Social services agencies typically use the term *intake* in lieu of *registration.* If a person cannot be scheduled for service, there is no access to those services. Once a person arrives for service, there is a registration or admitting process to gather important information about identification, clinical or behavioral reason for care, financial arrangements, and family or contact persons such as primary caseworker or primary physician.

A good strategy to address here is how this process will happen as settings for intake change and expand. Is the information system capable of adapting and being incorporated into a community-wide setting for health? There has traditionally been no coordination of care among healthcare, health, and social services providers within a community, so registration for one has no relationship to registration for services with the others.

Client/Patient Rights and Satisfaction

This dimension addresses the rights of individual patients, their families, and the public and the code of behavior of a provider organization. Organizational ethics includes religious beliefs and expectations and acceptable and unacceptable behaviors expected to promulgate the doctrine. Catholic hospitals, for example, are ethically opposed to abortions and do not support activities leading to voluntary termination of pregnancy, such as referral counseling.

Many of the external bodies focus heavily on the rights of patients and clients. They ask organizations to consider factors such as participation in decision making, informed consent, and cultural and spiritual values. Continuous assessment of the relationships and the members' ability to exercise these rights is important. The focus here is to help organizations conduct business relationships with their population and the public in an ethical manner by establishing a code of ethics, putting processes in place that ensure that member rights will be respected, defining the members' responsibilities in their own care, and establishing a mechanism for resolving problems. The mechanism for resolving problems is especially important to ensure that clinical, and not financial, data drive the behaviors, practices, and decisions of providers and the system.

Another area of focus for external review organizations is the system by which organizations or systems solicit patient input, measure their satisfaction, and build and maintain relationships with the population served. Measurement is required in order to assess how well the patient care delivery system is meeting these needs. This measurement is required beyond the patient. It includes evaluating and improving relationships with other customers and looking at referrals, loyalty, retention of market share, and satisfaction compared to the competition.

Organizational ethics are particularly challenged during the transition from fee for service, through fixed price per episode, to capitated reimbursement. Healthcare and social services organizations, physicians, and other providers find themselves in the ethical dilemma of conflicting actions to optimize the health of a population, minimize costs directly to the patients, minimize costs to payers and businesses, and maximize their own revenues. In most cases, financial viability depends on some balance among these actions, causing ethical and emotional tensions.

Prevention and Education

Health, healthcare, and social services systems are now asked to expand focus on the patient/clients to include educational needs, including self-care, specific information about a disease or illness, and even broader health promotion and disease prevention within the population served by the network. These external expectations reflect a process of assessing what patients and families need to know to manage their illness adequately, how they learn, and language and other barriers to learning or communication. Another important consideration is the information needed following discharge as the movement occurs from one setting to another one.

Many of these external review bodies have started to address the role of prevention in care delivery. Their common goal is to assist health and social services organizations in addressing the maintenance of health and preventing acute diseases, and avoiding, delaying, or providing rehabilitative resources used for chronic, degenerative, and disabling conditions.

Managing the Environment: Building, People, and Process Safety

The goal within this dimension, whether from OSHA, the JCAHO, or others, is to provide and maintain a safe and functional environment for staff and patients and other clients. The environment can refer to the entire scope of places where patients enter the system and consists of three basic components: the buildings where treatment is delivered, the equipment used, and people. Safe and effective management of the environment includes reducing and controlling environmental hazards, eliminating or controlling risks, preventing accidents or injuries, and maintaining safe, healthy conditions for everyone.

The processes to maintain a safe and healthy environment are a challenge to all organizations, especially in healthcare, from two perspectives. The first is maintaining a safe environment similar to any other organization, which can be difficult for those with poor social and environmental conditions and few resources. The second challenge, due to the communicable nature of many health problems, is to maintain an environment that minimizes the transmission of communicable health problems among patients, family members and friends, and staff. This dimension requires additional housekeeping and environmental services, as well as special precautions in handling hazardous materials used to treat patients. Reasonable precautions regarding infections, however, apply to all settings, particularly regarding serious infections such as the human immunodeficiency virus (HIV) or hepatitis. Most healthcare organizations have adopted uniform body substance precautions that possibly should be applied in all settings.

Assessment and Integration

Since these external expectations identify the important functions and processes an organization performs, they can be used to craft and design better delivery systems. The benefit of self-evaluation is that it allows you to understand the strengths of your organization. This information is what you have to bring to the negotiation table. The evaluation provides an accurate, clear picture of your strengths and weaknesses, what you have to offer, and what you strategically need to have, and do, to survive.

Initial actions toward integration will be enhanced by talking a common language. Discussions revolving around these functions provide an overview or macro-level insight into the work that needs to be done to integrate the system. It allows each component or organization within the system some understanding of individual and system goals. This is in addition to the micro- or organizational-level understanding by each organization on how well it currently performs these necessary functions. How well each organization in the system performs will be crucial information in determining levels of quality, cost, customer satisfaction, and ultimately survival. Assessment utilizing a common set of criteria can serve as the initial action plan to integrating a healthcare delivery system. These expectations are something organizations have in common and provide a neutral framework from which productive discussion and actions can emerge. The discussion about how each organization approaches the various functions provides a forum where emotion and ego are minimized and the basic discussion revolves around quality and improvement. These dimensions force important discussion about mission, vision, values, and customers. The focus is heavily on the role of leadership, information, and the actions that leaders within the health system need to take. It does not matter what type of health, healthcare, or social services organizations are involved in the evaluation; negotiating and assessing the system from a common base of expectations will improve the chance of obtaining the information needed to make appropriate decisions.

Assessment and the Marketplace

Purchasers are emerging in several different forms. The traditional third-party insurance companies, the self-insured organizations, the government, managed care organizations, and others are all examples. When a purchaser makes a choice to use one organization instead of another, the decision will most likely be based on both quality and cost. Consumers of health services are being educated on how their purchasers choose the organizations. In order for managed care organizations and other purchasers to "sell" prospective payments and managed care, the consumers must be assured that quality will not suffer during this shift. Quality is becoming a given for survival. Most companies understand regulations and accreditation, and that is why accreditation is used as an indicator of quality.

The form of healthcare delivery is changing. The processes are being integrated into a system or continuum of care serving a population, not just a patient. There are increasing demands by consumers, purchasers, and payers for accountability and accessibility. Low cost and high quality are both expectations for the evolving health system, while at the same time pressures are being applied for regulatory and financing reform. Commonly the systems that are evolving are much more elaborate, with varying degrees of complexity and relationships. Coordination and cooperation are keys to the success of these systems. External expectations identify organizational processes that must be improved in order to integrate successfully. These same expectations must be met as a matter of doing business. Most systems are currently somewhere in a transition period from individual organizations to partners in a system. This set of common dimensions and criteria reflects a national consensus on quality and performance expectations for health and social services organizations. That is why we are calling for a set of expectations for virtual systems, which aids these organizations in collaboration rather than sets each one up as distinct. The expectations serve as excellent guides to good management practices and excellent clinical care. In addition to being well managed and providing high-quality patient care, these systems are being held up to public scrutiny and accountability as never before. Expectations have increased even as the complexity of the organizations and systems has increased. The social services and healthcare systems in a community must rise to meet this challenge.

Successful systems are those that can master both the clinical and management functions. External expectations can highlight and serve as guidelines for evaluating a system and offer areas for improvement. The evolving systems are complex systems of interrelated and interdependent processes. The important processes identified in external expectations are areas that can serve as a framework and should be addressed in all components of the system; they include quality, with monitoring, measurement, and improvement throughout the system. These criteria provide a starting point for evaluation and discussion.

CHAPTER SIX

INCORPORATING EXTERNAL EXPECTATIONS INTO INTERNAL PROCESSES

Now that you have an understanding of the use of external expectations in helping the system integrate and succeed, what is the best approach to analyze your organization, a new or existing set of organizations, or competitors? In this chapter, we will relate the scope-of-services tool, introduced in Chapter Four, to the common dimensions identified by external review bodies. In an effort to demonstrate once again that allocating resources for each review organization is a waste of time and money, we have reviewed and consolidated dozens of external expectations and distilled them into nine dimensions, with multiple criteria under each dimension. These dimensions, along with the criteria applicable to assessing performance, are listed in Figure 6.1. The intention is to provide the organization and components a common framework for assessment and evaluation and then to cause your organization and your partners to begin to incorporate all of the various external expectations each individual organization is subjected to into an integrated, process-oriented approach.

The Approach

We have seen that although there are multiple review, licensure, accrediting, and recognition expectations to contend with, there is a relatively small and succinct set of common expectations that overlap throughout all of them. These

FIGURE 6.1. ANALYSIS OF EXTERNAL EXPECTATIONS.

Major Section		Important Criteria	Criteria Scores Score (0–4, 4 = high)			Comments
Number	Name	Description	Your Organization			
			d	e	f	
a		b	c			

Scoring of criteria (fill in the boxed areas):
0 = No competence, coverage, or success, or not applicable
1 = Minimal competence, coverage, or success
2 = Partial competence, coverage, or success
3 = Significant competence, coverage, or success
4 = Substantial competence, coverage, or success

2   External Expectations

Leadership and Governance
Communications
Community involvement
Establishment and alignment:
  Mission
  Vision
  Values
  Critical success factors
  Goals
  Key processes
  Key indicators
  Feedback
Other priority criteria, identified
  by organization

Strategic Planning
Current financial status
Financial objectives
Customer requirements
Customer expectations
Resource allocation
Staff requirements

FIGURE 6.1. ANALYSIS OF EXTERNAL EXPECTATIONS. (continued)

Major Section		Important Criteria				Criteria Scores Score (0–4, 4 = high)			Comments
Number	Name	Description				Your Organization			
			a	b	c	d	e	f	

Staff expectations
Environmental changes
Organizational capabilities
Future risks
New services
Services redesign
Other priority criteria, identified by organization

Human Resources Management and Development
Alignment with organizational goals
Staffing (right number of people)
Job categories (right kind of people)
Skills assessments
Competency evaluations
Credentialing clinicians
Employee satisfaction
Employee development
Continuing education
Training
Other priority criteria, identified by organization

FIGURE 6.1. ANALYSIS OF EXTERNAL EXPECTATIONS. (continued)

Major Section		Important Criteria			Criteria Scores			Comments
Number	Name	Description			Score (0–4, 4 = high)			
					Your Organization			
a			b	c	d	e	f	

Process and Quality Improvement
Input from:
  Clients/patients
  Staff
  Clinicians
  Vendors/contractors
Geographic settings identified
Assess external environment
Identify all important processes
Develop measures for processes
Monitor processes
Evaluate data
Implement/improve processes
Other priority criteria, identified
  by organization

**Information Planning
and Management**
Needs assessment
Identify:
  External data requirements
  Internal data needs
Solicit input from:
  Clinicians
  Staff
  Suppliers
  Providers

FIGURE 6.1.   ANALYSIS OF EXTERNAL EXPECTATIONS. (continued)

Major Section		Important Criteria	Criteria Scores Score (0–4, 4 = high)				Comments
Number	Name	Description		Your Organization			
	a		b	c	d	e	f	

Information needs for:
Any client/patient
Any time they enter system
Any place they enter system
Plan from needs assessment
Address:
  Confidentiality
  Security
  Accessibility
Access to literature
Coordinate medical record:
  Numbering system
  Patient-level data requirements
  System access
  Uniformity
  Standardize operational definitions
Other priority criteria, identified
  by organization

Continuum of Care
Patient/client care
Integrate:
  Admission/entry/intake
  Registration
  Scheduling coordination
Assessment:
  Care needs are assessed at all sites

FIGURE 6.1. ANALYSIS OF EXTERNAL EXPECTATIONS. (continued)

Major Section		Important Criteria				Criteria Scores Score (0–4, 4 = high)			Comments
Number	Name	Description	a	b	c	Your Organization d	e	f	
		Assessments are appropriate to site							
		Level of care provided is appropriate							
		Services are appropriate							
		Services are timely							
		Integrated treatment/services:							
		Communication between sites							
		Communication between practitioners							
		Transfer between sites is seamless							
		Discharge							
		Follow-up/aftercare							
		Population needs assessment:							
		Sufficient disciplines and specialists available							
		Appropriate services available							
		Appropriate levels of care available							
		New service design							
		Existing service evaluation							
		Measure processes							
		Improve service/processes							
		Incorporate support services							
		Identify community resources							
		Incorporate community services							
		Incorporate administrative support services							
		Population eligibility for services							
		Mechanism to communicate services							

FIGURE 6.1.   ANALYSIS OF EXTERNAL EXPECTATIONS. (continued)

Major Section		Important Criteria	Criteria Scores Score (0–4, 4 = high)				Comments
Number	Name	Description	Your Organization				
	a		b	c	d	e	f	

Systemwide access issues:
Transportation
Hours of service
Cultural barriers
Language barriers
Other priority criteria, identified
by organization

Client/Patient Rights and Satisfaction
Informed consent
Client/patient participate
in decisions
Recognition of:
Cultural differences and values
Religious practices and values
Patient/client input gathered
Patient/client satisfaction measured
Information used for process/
service improvement
Ethics/behavior code addressing:
Conflicts
Incentives
Decision making
Mechanism for resolving problems

FIGURE 6.1. ANALYSIS OF EXTERNAL EXPECTATIONS. (continued)

Major Section		Important Criteria	Criteria Scores Score (0–4, 4 = high)				Comments
Number	Name	Description	Your Organization				
a		b	c	d	e	f	

Other priority criteria, identified by organization

Prevention and Education
Education needs determined for:
Patient/client
Family/support
Community
At-risk populations
Population health promotion is:
Assessed
Planned
Implemented
Disease prevention is:
Identified
Assessed
Planned
Community resources are included
Barriers are identified:
Economic
Family/social
Language
Access
Aftercare is:
Assessed
Tracked
Monitored

FIGURE 6.1. ANALYSIS OF EXTERNAL EXPECTATIONS. (continued)

Major Section		Important Criteria	Criteria Scores Score (0–4, 4 = high)			Comments
Number	Name	Description	Your Organization			
	a		b	c		
				d	e	f	

Other priority criteria, identified by organization

Managing the Environment
Plan development incorporates:
  Building
  People
  Equipment
Process for monitoring:
  Safety and infection control
  Controlling environmental hazards
  Address risks
  Accident prevention
  Building codes
  Fire codes
  Disaster plans
  Local requirements
  State requirements
  Federal requirements
  Implement improvements
Other priority criteria, identified by organization

crosswalks helped us identify nine dimensions of performance that organizations that are assessing themselves, potential partners, the system, or competitors will want to evaluate. This tool is used to organize that information for analysis. Remember that while gathering the information and performing the analysis, you are looking for these areas:

- Gaps in specific areas where external expectations will apply
- Duplication of each function or process between these various organizations or services that may allow for some consolidation of information
- Conflicts among services, functions, or standards
- Appropriateness in relation to all the various organizations subject to external review and among components of your system

    We suggest beginning with the creation of a list of all external bodies with which your network is involved. A matrix is a good tool to consider at this level. Compiling this list may not be easy because larger, more complex entities can be reviewed by over forty different external bodies. This is also the time to become familiar with potential or added components, such as a child welfare agency or another social services program. Although we demonstrated with the crosswalk that the processes are essentially the same, the language can be very different and may require time to clarify definitions for the people doing the work. As mergers continue and integration evolves, if you have added other services or organizations, then you probably have added other external expectations. Compiling a list of various external and accrediting bodies, and their area of focus, is a huge project. Again, we refer you to the *Lexikon* (Joint Commission on Accreditation of Healthcare Organizations, 1994). This first step should be conducted by a representative group of health and social services clinical leaders who are familiar with their respective organizations or services.

    One remarkable resource you probably already have within your organization and throughout the network is a group of highly trained people whose job it has been to get you through these inspections. Many organizations have a department of quality or survey preparation. These people have attended seminars, read and memorized the expectations, and thus have an excellent knowledge base. It would be helpful to gather these people together and determine the expectations everyone has to meet, if there are any problems, or whether there are particular processes that one part of the organization does exceptionally well.

    We suggest looking for areas of overlap among the external expectations as well as major gaps related to newly added components. Assess for duplication or overarching standards that can be incorporated into an entire network and still address individual expectations. For example, how you manage the flow of patient information is an expectation of several major external bodies and also translates into more focused expectations specific to service areas in your system. This same process is one of the major criteria in most state and national quality awards. As you begin to go through these expectations and lay them out, you will see that many of the state and national quality awards reflect the expectations within the external bodies you must already meet.

A horizontal and vertical analysis will help move you from the initial matrix list of components and external expectations into a more detailed approach to evaluating and assessing the component against the dimensions. This process will help your organization or system with the benefit of graphically illustrating for any external reviewers or surveyors just how their set of expectations is incorporated into the organization or systems processes.

The second step in the analysis, after you have compiled an exhaustive list, is to familiarize yourselves with the area of focus. We suggest prioritizing them in the following manner:

1. The external expectations that must be met for legal or licensing expectations
2. The expectations that must be met for funding or reimbursement
3. The expectations of specific groups of clinicians
4. Desirable certifications or awards, such as state or national quality awards, for gaining a competitive advantage

Different types of people may be involved in the analysis of external expectations at varying levels of assessment:

- Internal leaders and staff who are knowledgeable about the subject of the analysis, especially in the initial gathering and prioritizing of information about various external expectations
- Staff internal to the organization but outside the division to be assessed
- Members of the organizations involved in the analysis, in a collaborative effort
- External reviewers or consultants
- Official reviewers

The Analysis Tool

A standardized tool provides a common format for people to use, the language and definitions can be agreed on prior to utilization, thus allowing less room for personal interpretations, and it does not require a reinvention of the wheel. We continue to use a common format for the analysis throughout the book. Of course, the analysis tools can and should always be tailored to the needs of your organization and the purposess of the analyses. Our sample tool for analyzing external expectations is illustrated in Figure 6.1.

Selected important criteria representing a consolidation of many external review expectations are listed under each of the nine dimensions. You will notice throughout this section that is there is great deal of overlap of criteria among dimensions. Although each dimension represents the processes your organization must do well, just as systems are full of complex interrelated processes, these dimensions and criteria are similarly interrelated. How well a system does in one dimension will affect how well it does in others. At the bottom of each list of criteria is space to add other priority criteria for your organization. Also note that there are blank columns, which are used for the combined analysis matrix described in Chapter Ten.

Columns d, e, and f are illustrated for scores related to each of the criteria. The score for your individual organization is entered in column d. Columns e and f are used for scores of your partner organizations or competitors. Additional columns could be added for more partners.

The following five-level scoring system is illustrated in Figure 6.1:

0   No competence, coverage, or success, or not applicable

1   Minimal competence, coverage, or success

2   Partial competence, coverage, or success

3   Significant competence, coverage, or success

4   Substantial competence, coverage, or success

Since there are likely to be differences among departments or services, you may choose to assign scores based on different approaches:

- Blended score. This approach considers all of the different units, and assigns a score based on a weighted average of all the units.
- Low score. This is the approach used by the JCAHO, when it assigns a poor score and recommendation for change based on finding poor performance in one area.
- Differences among areas may also be indicated, since this gives an indication of variation within your organization or between your organization and those with which you are comparing.

Although each organization will place different weights or values on different dimensions, depending on their own system, a discussion of possible ideal scores or ranges might be of value. For the initial analysis for your own organization, a simple scoring system is recommended. The question you are attempting to answer is how well your organization, partners (potential or existing), or competitors are addressing the different external expectations.

There are scoring systems attached to most of the external expectations discussed and consolidated here. One suggestion is to incorporate a scoring system such as that of the Baldrige Award. A perfect score in the Baldrige Award process is 1000. A score of 700 will put an organization in the winner's category for the award, so shooting for a score of 500 to 700 would be an appropriate initial goal for your system. The JCAHO has a simpler scoring system, with a total of 100 points.

The approaches will vary, and this is the advantage of the tool. It is standardized but provides flexibility to meet different needs. The tool also provides criteria and interpretation for each dimension, as well as how to ascertain the assessment score for each. Again, we suggest tailoring this tool to fit your priorities. If, for example, you choose to place much more attention on information management and planning, either across the system or within a component, look at and weight the criterion. Expand on the criterion so you are able to capture the appropriate level of detail for evaluation of the dimension.

A column in Figure 6.1 is provided for comments. Comments are most helpful if they address the need for clarification, the need for more detailed analyses, and, in particular, the type of comparative information that is desired.

Analyses of External Expectations

This section provides criteria and some general questions designed to help the person doing the assessment determine a score related to each of the nine dimensions of external expectations. In a more detailed analysis, more precise and pertinent questions need to be developed to ascertain the score. Recall that a key purpose of these analyses is to expand your thinking beyond each individual service, department, or organization into process improvement for the entire system. Ascertaining the scores includes several approaches, similar to the external review processes. Interview members of services, departments, organizations, leadership, and other individuals who understand the process criteria you are scoring. Quantitative information should always be documented and accessible. Many of these criteria will be overlapping. Detailed discussion for designing your scoring approaches should also include people you have identified in your system who are most familiar with various expectations. Again, it will be helpful if gaps, duplications, conflicts, issues, or more data needed are identified in the comments section.

Leadership and Governance

Ten leadership and governance criteria are listed in Figure 6.1: communications; community involvement; and establishment and alignment of mission, vision, values, critical success factors, goals, key processes, key indicators, and feedback. For a general overview analysis, the following questions or documentation can be used to assign a criterion score:

- How is the community involved with the organization or system? In what ways and in what areas? How is input sought? What examples are there that this input is being used within the system, organization, or department?
- How would you demonstrate that the organization or system has established mission, vision, values, goals, and critical success factors? How are these communicated throughout the system?
- How were key processes identified? What are the key indicators? What is the scope of the processes or indicators (service or setting specific? systemwide?)? Who was involved in their development?
- How is feedback reported? To whom? What do these people do with the feedback?

Common problem areas include alignment of customer expectations, financial planning, and prioritization. Another problem is that many individual organizations can demonstrate success within a dimension or set of criteria, but when this is evaluated over the system, many of the linkages fall apart or duplication of efforts is noted.

Strategic Planning

The twelve criteria listed for strategic planning are current financial status, financial objectives, customer requirements, customer expectations, resource allocation, staff requirements, staff expectations, environmental changes, organizational capabilities, future risks, new services, and services redesign. The following questions will help in assigning scores to the criteria:

- How were financial objectives determined? How do these objectives fit in with the strategic plan? How are they incorporated into quality planning?
- What are the specific customer requirements you are focusing on? How were they developed? How do they fit into the financial objectives?
- What are specific staff requirements? How were these developed? How do they fit into the financial and quality planning?

- How are the financial objectives and customer and staff requirements prioritized? By whom?
- What population information do you have about the community? How do you receive and share information with community resources and other organizations?
- How do you assess your organization's existing capabilities? How do you plan for changes within the population and environment?
- How have specific services been developed or changed with the integration? What is the focus?

Common problems with strategic planning occur if the planning has been done by only a few people, with minimal input from clinicians, staff, and the community. Lack of information about the population, trends, and other community resources will cause the process to fall short. When this occurs, the score would be low. The comments section may be used to identify further information.

Human Resources Management and Development

Criteria to consider for this dimension are alignment of organizational goals, staffing (the right number of people), job categories (the right kind of people), skills assessments, competency evaluations, credentialing clinicians, employee satisfaction, employee development, continuing education, and training. Questions for scoring purposes include the following:

- What information did you use to determine the job categories?
- How can you demonstrate competency across the system? How is it determined for different services or departments or other areas? How are the criteria communicated?
- How is employee satisfaction assessed? Can you give an example of employee feedback that has led to an improvement in an area?
- What type and how many clinicians do you employ? How are their credentials monitored and updated? By whom? How can this information be accessed if needed?
- How do all staff, clinical and support, give and receive feedback regarding system issues to the leadership? Can you cite an example?

Process and Quality Improvement

Eleven criteria are listed in process and quality improvement: client/patient input, staff input, clinician input, vendor/contractor input, geographic settings identified, assessment of external environment, compilation of processes, measurements for processes, monitoring of process improvement, evaluation of process improvements, and implementation of actions to improve. This section also provides the framework for evaluating and improving all of the other processes identified in the nine dimensions. Therefore, although the questions suggested below are broad, there should be evaluation at detailed levels throughout the assessment process to incorporate these criteria:

- To what degree can your system demonstrate success in this dimension?
- How are customers and employees involved? How are they informed of changes?
- How many and what data have you collected and analyzed about your processes?
- What type of information did you get from the data collected? How was it used? What was identified for your organization or system in terms of strengths and areas for improvement?
- What types of approaches do you use to measure, assess, and improve in different areas?

Information Planning and Management

There are twenty criteria in Figure 6.1: needs assessment; identification of external data requirements and internal data needs; input from clinicians, staff, suppliers, and providers; information needs for any client/patient, at any time they enter the system and any place they enter the system; development of a plan based on a needs assessment; provision of confidentiality, security, and accessibility; access to the literature; and coordination of all types of existing records: numbering system, patient-level data requirements, system access, uniformity, and standardization of operational definitions. The topic is so broad that we suggest following the outline for levels of information and its use with the system. The first questions should be

- What is your approach for identifying and developing a plan for managing information required by the system?
- How many different patient identifiers are used (medical record numbers, case records, social security numbers, and so forth)? How is this issue addressed throughout the system, at any point of entry, and how is it communicated to the appropriate caregivers? How are these identifiers reconciled?
- What examples do you have to demonstrate the use of aggregate information in developing treatment options and plans? Has there been any focus on treatments and groups outside medical diagnosis–related groups (DRGs)?

- What are the characteristics of your community? What role do you play in providing required services to it?
- Are there gaps in the information processes? Are there duplications?

Continuum of Care

This dimension, with thirty-three identified criteria, is by far the largest and most inclusive dimension. It includes the perspective from the individual client/patient as well as the population served. In assessing the organization or system against this criterion, you may want to break it down initially and evaluate each criterion both horizontally across the system and vertically down each component. This approach will make assessment more manageable and still produce a comprehensive assessment. Criteria for evaluation and scoring include assessment and intake, criteria for treatment, and discharge planning. The system review includes the design of new services, evaluation of existing services, implementation of measurement and improvement of process methodology, types of support services utilized, identification and utilization of community resources, incorporation of administrative services into service delivery, assessment of population eligibility for services, establishment of mechanisms to communicate service, and establishment of mechanisms to ensure access across the entire system, including how the system addresses transportation, hours of service, cultural barriers, language barriers, physical barriers, integration of admission processes, registration processes, and scheduling systems. Following are some questions to address:

- Does the system formally address potential areas to improve in this dimension?
- What are the priorities in terms of the criteria listed above? Why?
- What quantity of data is collected and analyzed within each criterion?
- How have you consolidated the data from various components in your system into a common set?
- Are there duplications of efforts or gaps identified?

Client/Patient Rights and Satisfaction

Eleven criteria are listed in Figure 6.1: ensuring informed consent; establishment of a process for client/patient participation in decision making; recognition of cultural differences and values; recognition of religious practices and values; inclusion of patient/client input; measurement of patient/client satisfaction; incorporation of information in quality process across the network; establishment of ethics and a behavior code addressing conflicts, incentives, and decision making; and establishment of a mechanism for resolving problems. General questions include the following:

- How do problems concerning clients' rights get reported?
- How do they get resolved?
- How do you measure satisfaction? How is satisfaction information collected?
- Are there gaps or duplicative efforts in this area?
- Who gets the information?
- Are there examples of improvements made as a result of information collected regarding patient satisfaction?
- What is the organization or systems approach to capturing information about various beliefs and cultures and their importance to caregivers?

Prevention and Education

The eighteen criteria encompass individual patients and the community served by the system. Educational needs are determined for the patient/client and for family members. Health promotion is assessed, planned, and implemented, and disease prevention is addressed and implemented. Barriers to learning are addressed, including language gaps, social and family issues, reading limitations, and sensory (hearing or seeing) difficulties. Follow-up and linkage with other resources within the community are assessed and provided for after discharge. Information provided to the leadership on prevention will be useful in long-term strategic planning for the system. Responsibility is shifting to systems of care for managing the health of a population. Prevention has potential long-term payoffs, but there has been little incentive in the past to justify buying into a long-term strategy because of the need for short-term gains. We do not know exactly what will force this issue eventually, but we believe that organizations that begin to incorporate strategic initiatives toward prevention will have less difficulty later in the evolution. The collaboration of health, social services, community resources, and any other potential population/community partners identified by visionary system leaders will make this process more successful. Scoring questions include the following:

- Does the organization or system have any recognized, budgeted services that address this focus?
- Has the organization or system achieved any measured success in this area?
- Have you identified what types of preventive services or education are required by the population you serve? How did you make this determination?
- Do you include any other community resources in this service?
- Are there duplications noted?

A higher score would be appropriate in this dimension if the system conducted a community needs assessment in order to determine that education and preventive services would be appropriate. The system leaders incorporate the outcomes of the needs assessment into strategic planning for the system and include quality improvement activities and ongoing evaluation of the preventive and education services.

Managing the Environment: Building, People, and Equipment Safety

There are fifteen criteria in Figure 6.1. For the entire network, there must be a plan in place for monitoring safety and infection; a mechanism to reduce and control environmental hazards; a mechanism to identify, control, and eliminate risks present in the environment; a plan to address accident prevention; a plan for infection control; and mechanisms to ensure that building and fire codes are met at the local, state, and national levels. The entire safety plan must be continually developed, monitored, and improved. Some important questions are these:

- Does the organization formally address this dimension? What has been the impact?
- How do you incorporate this approach into nontraditional settings?
- What education and training do you provide to personnel within these areas?
- How is this entire dimension continually assessed?
- Are there gaps in this area?

An example of a system approach to addressing this dimension would be evidence that each component site addresses, through planning and training, issues such as security, safety, accessibility, equipment management, space adequacy, trained personnel and other safety measures.

Completing the Assessment

In the comments section, it may be helpful to identify the major gaps, duplications, conflicts, and issues so they can be addressed. For the purposes of the general overview analysis, the following questions can be used for the dimensions/criteria and assigning a criteria score:

- How well does the organization perform this function? Whereas an organization may not offer some services, with some foci, or in some settings, most organizations will have all these processes. The question is how well those processes meet the needs of patients and other customers.

- Are there inappropriate duplications or conflicts? Mergers create huge conflicts among these basic processes.
- Do staff understand and use the process well?
- Are there complaints about the process? If so, the score will be lower.
- How many positive versus negative examples can be cited about the process?
- What data exist to demonstrate that the process is successful?

Steps to Using the Analysis Tool

Now that you understand the approach to conducting the analysis and how to use the tool, the next step is for your organization to tailor both the process and the tool to meet your requirements. We suggest answering the following questions to help you achieve this step:

1. What is the purpose of the analysis?

    Are you analyzing a single organization?

    Are you analyzing an existing network?

    Are you analyzing potential partners for the network?

    Are you comparing your organization or network against competitors, network partners, or both?

2. What is the level of detail you need for your current assessment?

    Is it simply an initial step toward meeting external expectations?

    Is it to evaluate a key dimension throughout the system?

    Is it an initial step to look at self-evaluation and what your organization can bring to the table or needs to survive?

    Is it a detailed, complete assessment for evaluating your strengths and weaknesses in order to identify marketable advantages or develop strategies toward differentiation between your system and your competitors?

3. After reviewing the criteria used in Figure 6.1, are there others to add, delete, or amend? Are there certain overarching functions, such as management of information, that are more important and therefore need to be much more detailed?

    Revise the tool to meet your requirements.

    Select the people to perform the assessment and the analysis.

    Review the analysis tool, the operational definitions of terms, and the scoring method with the people involved in this process.

Horizontal and Vertical Approaches

These two different approaches are helpful for using external expectations for analyzing your current organization or potential partners. Exhaustive evaluation of all the dimensions for external review organizations, within all settings, can be unreasonably time-consuming if approached in a disorganized, piecemeal manner. More focused analyses along the dimensions instead of by review requirements may be more appropriate.

To look at the external expectation dimensions or criteria horizontally, you select one and look across the taxonomy at all the settings that might apply to that dimension and to all external organizations that may require this dimension. This approach is illustrated in Figure 6.2. For example, you could begin with a particular dimension, such as information planning and management, and then look across at the external review bodies that require some aspect of this within their expectations and if you have addressed this dimension and criterion in all of your system's settings. When preparing for assessment or a review, use the horizontal analysis approach to address expectations or standards associated with that review program.

Another approach is to look vertically at one particular setting, perhaps a newly acquired one. You can assess the setting against dimensions addressed in it, as well as against any external review organizations that may focus on this particular setting. This is also illustrated in Figure 6.2. For example, say you have recently added a long-term-care facility to your system. Assessment of the nine dimensions, as well as what external review organizations apply, will give you a

FIGURE 6.2. HORIZONTAL AND VERTICAL INTEGRATION OF EXTERNAL EXPECTATIONS.

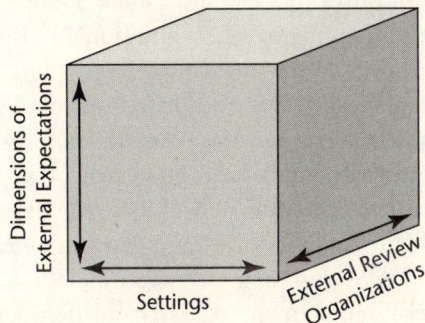

good idea of where the gaps may be and what the focus should be toward integration and improvement efforts.

Levels of Analysis

The overview analysis typically is based primarily on perceptions and readily available data. This approach provides a snapshot assessment of the organization. Minimal effort is spent collecting data for the overview analysis. This level of assessment provides broad, qualitative results.

It is easy to believe that your processes are all working well, but unless you or another unbiased evaluator test it against proven, defined criteria, how do you really know? What is said and what can be documented may be very surprising. When you want to quantify the processes and demonstrate the level of quality, you may fall short. If this may be the case, the organization or system leaders want to move into a much finer level of detail, an effort that requires more time, resources, and sometimes outside assistance to assess and document in detail.

If you decide to conduct detailed and quantitatively based analyses concerning selected topics, you can nevertheless use the same analysis tool. The difference now is that the scores entered for each criteria are based on more quantitative data. Chapter Ten explains a consolidated audit analysis tool.

Actions Based on Analyses

The external expectations analyses, using Figures 6.1 and 6.2, will lead to the identification of a number of actions. Following are examples of the types of actions and additional analyses that your organization may undertake:

- Identify opportunities. By the time you have finished the analyses using the tool, you will almost certainly identify some opportunities for improvement—for more integration of processes, additional settings, or different levels of key processes.
- Identify duplications. You need to identify duplications within your system or among your organization and potential partners. Some of these duplications will be unnecessary and costly; others will be appropriate.
- Determine the appropriateness of external review expectations. Appropriateness is difficult to evaluate because different people have different perceptions of appropriateness. Basing this determination on the healthcare, health, and social services requirements of the populations served is an effective way to an-

swer the appropriateness question. You can begin by prioritizing the expectations by those that you must meet for legal reasons, those you must meet for funding or reimbursement reasons, and so forth.

- Identify issues and conflicts for resolution. Assess the compatibility of potential partners to identify what issues, conflicts, strengths, or weaknesses are evident.
- Identify and prioritize areas for more detailed analyses.
- Identify approaches to fill gaps. Who might be responsible for this: an individual or a team?
- Identify approaches to eliminate inappropriate duplication.
- Develop collaborations to maximize quality and cost-effectiveness.
- Determine how to incorporate the results into strategic initiatives.
- Identify approaches to fill gaps.
- Identify approaches to eliminate inappropriate duplication.

The results of the analyses related to consolidation of external expectations are key inputs to assessing or developing an integrated health system. The scores for the specific criteria included in Figure 6.1 may be used in two different ways. If you choose to do the combined analyses at a more general level, then the specific results from the analyses discussed in this chapter serve as the basis for the scores assigned in the combined analysis table. Alternatively, the more specific criteria in Figure 6.1 may be incorporated into a more detailed spreadsheet, replacing Figure 10.1.

CHAPTER SEVEN

UNDERSTANDING THE INTERNAL ORGANIZATIONAL CLIMATE

The taxonomy provides a classification model for analyzing a broad range of potential components of health and social services. Yet we find that the design and implementation process has additional, less tangible but extremely powerful expectations in terms of the soft side of organizations: the human dimension and internal organizational climate that so pervasively affect all implementation and outcome (see Figure 7.1).

Corporate culture and the human systems of organizations can best be defined as the expression of values and beliefs through behaviors and actions. Each organization, regardless of size or purpose, possesses a culture: a perceivable, assessable, "way we do business here" approach to work (Kotter and Heskett, 1992). High-performance organizations exhibit predictable common traits in the assessment of internal organizational climate and this climate's demonstration in human systems. These common traits and their corollaries in external expectations can be assessed, gaps determined, and appropriate action strategies to address shortfalls or overused strengths implemented.

During one of our early consultative engagements for a major medical center, we were charged with assessing the causative factors for "low morale" in middle management. A core part of the assessment was one-on-one interviews with the management staff (the best diagnosis can frequently be obtained from asking the patient what is wrong!). In the middle of an interview with a nursing supervisor, one of us asked the standard question, "What do you like least about your job?"

FIGURE 7.1 INTERNAL ORGANIZATIONAL CLIMATE.

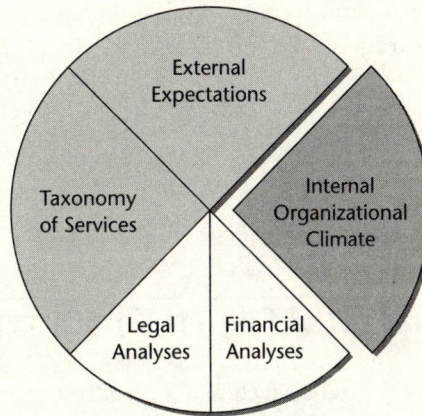

The astounding reply was, "If it weren't for physicians and patients, this would be a great place to work." As incongruous and humorous as the reply appears, it illustrates the phenomenal importance of mastering the art of human systems management and the degree to which this art is neglected and discounted. Nothing is more of a giant killer than failure to respect the power of the soft side.

The creation of effective and efficient VIHS delivery systems is fundamentally about initiating and managing the impact of massive organizational change on people. Failure to address adequately the always dynamic, frequently unpredictable, and often frustrating forces of rapid change spells organizational doom. High-performance organizations have mastered the art of change management, at least temporarily. But without constant renewal and tending, the practice of change management can be lost or made less effective.

All organizations, whether hospitals, health systems, social services agencies, or manufacturing plants, possess common characteristics and structures. Someone leads, someone follows; something is produced, either goods or services; purchases are made and products sold; hiring, supervising, and firing occur. We term the efforts to control these various organizational processes *management* and diligently attempt to exert control and predictability in the complex systems that have evolved.

The structure of the proposed delivery system, whether owned and operated by one sponsoring corporation or assembled from an amalgam of affiliations and federations, is of less importance. Of primary significance is the outcome of the management processes that can successfully guide the virtual system through the number of challenges and changes that await in healthcare.

The field of organizational development puts a focus on the theories and technologies of managing organizational change. As we attempt to build larger and more effective organizations, virtual or directly owned, we grapple with how to predict, encourage, and control change. Much of what anyone in a management position, from chief executive to line foreman, does is to manage the impact of change. Thus, the common attributes of highly effective organizations are perceptible regardless of structure, function, or purpose of the organization. Each of the core attributes analyzed in this template is fundamentally dependent on effective human systems and change management.

Internal Organizational Climate Characteristics

The internal organizational climate is a reflection of at least eight characteristics. Let us look at each in turn.

Structural Fluidity

One of the key characteristics of high-performing organizations is the ease with which the structure and shape of the organization mold and adapt to mandates for change. This property is best described as structural fluidity. Fluidity is demonstrated by the organization's consistent evolution toward the most effective and efficient delivery model for the times, with a focus on the best fit of people, position, and strategy. Organizational rigidity, commonly characterized by worship of an organizational chart and the operational hierarchy that it represents, is largely absent in high-performance organizations. In fact, truly effective organizations are constantly realigning people to reflect the needs of the environment. It can be said that these organizations use an "etch-a-sketch" organizational chart, and the standing joke we have heard is, "If my boss calls, get her name."

This structural fluidity supports and encourages what Peters and Waterman (1982) call a "bias for action": any action is preferred over inaction. In the face of organizational challenge, there is an implicit cultural value to doing something—almost anything. An inclination for action may mean the organization frequently stumbles but more frequently triumphs. The prevailing wisdom in these highly successful organizations is that failure to act is certain to mean defeat; thus action, even if sometimes the wrong action, slants the odds of success toward the actor. The expression "You miss 100 percent of the shots you don't take" is adhered to in high-performance organizations.

This organizational fluidity necessitates high turnover in point man positions. The point man position leads the charge in a particular area. For the organization to move quickly and with flexibility from one challenge to another, different people, with different talents, must take "point" on each issue. Rarely do high-performance organizations exhibit overreliance on one key person, though many are led by and associated with a "big" personality, such as Jack Welch at General Electric, who has led the company through phenomenally successful growth and change. Most very successful organizations exhibit fluidity in leadership through movement of key people and significant responsibility.

This rotation of talent lends the institution multiple points of view, with identified experts in various environments leading the charge, and develops a broad base of skilled people. It also distributes responsibility and accountability more equitably around the organization, allowing leaders to regroup away from the hot seat and revitalize, whether in the wake of success or failure. The distribution of leadership responsibility also increases participants' support for and endorsement of change. Each stakeholder knows that his or her turn at the helm is coming, and when it does, each will need the support of colleagues. Thus the rule of reciprocity gently applies.

High-performance organizations also exhibit structural fluidity through the existence of a spectrum of collaborations, from formally designated teams to spontaneous groupings of disparate individuals, that lead from innovation to implementation. The irreverence for rigid organizational structure is reinforced through "outbreaks" of teams and task forces that take charge of particular projects. Membership on these task forces and teams floats according to the team's mission and needs. Thus, there is a constant mixing—an ebb and flow of personality that not only contributes talents appropriate to project needs but also enhances and supports organizational communications as people come to know and value one another.

High-performance and fluid organizations put increased focus on customers and outcomes—the tangible, measurable, results of activity. In so-called silo-based organizations, with rigid internal divisions and departments, results can be measured against departmental objectives and budgets. Since there are few, if any, silos within the fluid organization, focusing on customers and outcomes is the only way in which activity can be evaluated. Outcome measurement provides the basis for this evaluation of organizational effectiveness.

Concomitant with outcome measurement is a decreased focus on process for the sake of process. The question asked of these organizations is not so much "how?" but "what, when, and where?" The process of development and implementation of strategy is not as important as the outcome of the development. This is not to imply that these organizations are cavalier in the design or implementation of new programs and products. To the contrary, deliberate research and development is the hallmark of innovation; nevertheless, deliberate design should not be confused with "analysis paralysis," a state of inactivity due to indecision and the constant need for more information before a decision can be made.

As one might expect, these organizations have a high formal and informal value for creativity. There is room to nurture the misfits, nonconformists, and others from outside their ken who provide the diversity that creativity demands. There is little emphasis on conforming to observable societal norms and more emphasis on adherence to less observable, but in many ways more important, organizational values. Creative people are welcomed and supported with the expectation that their creativity contributes to organizational outcomes. This is not to say that there is not considerable tension between the "creatives" and those who do the daily production of products or provision of customer services. However, this tension is managed for optimal performance.

In addition, highly fluid organizations evidence a great value for inquisitiveness. Knowledge and learning are constantly honored and supported, through formal training, individual extracurricular learning, research and development activities, ad hoc research groups ("hunting parties"), informal groups ("skunkworks"), and other less formalized methodologies. Continuous learning and "outside the box" thinking is not only welcome but expected.

This type of corporate culture recognizes that success does not come without risk, and thus failures occur. In traditional organizations, failures are typically punished and certainly not reinforced. Indeed, security, stability, and low-risk behavior are frequently rewarded in less functional organizations. But highly fluid structures support acceptable risk and expect occasional failures. Such organizations use failure as an opportunity to study what might work and, just as important, what does not work. In fact, high-performance organizations view safety and comfort as high-risk strategies, believing that complacency leads to a focus on keeping the status quo, a culture of rigidity, and the inability to compete. Low risk *is* high risk! (See Figure 7.2.)

High-performance organizations further express their fluidity through the pursuit of stretch goals: expectations that push an organization, team, or individual significantly beyond acknowledged capabilities. Organizational goals are visionary and exhibit the grand scheme of ambitious thinking, with little room for small accomplishment. These stretch goals, and the time and training for goal achievement, are expected and continually recognized. Thus, organizational heroes and myths—stories told within the organization to illustrate corporate values—celebrate successful venturing in the organization and out of it. The people who are talked about and honored in organizational legends reflect achievements in venturing forth and supporting the organization's grand vision. One need only look at the mythology that has arisen around Nordstrom to see the powerful impact of the hero's myth (Spector and McCarthy, 1995).

FIGURE 7.2.   STRUCTURAL FLUIDITY ALIGNED
WITH EXTERNAL EXPECTATIONS.

Attributes	Dimensions
Bias for action	Leadership/process and quality improvement
Turnover in point man position	Leadership and governance
Teams, task forces, and skunkworks	Strategic planning
Customer focus	Client/patient rights and satisfaction
Outcome measurement	Information management/process and quality improvement
Creativity valued	Strategic planning
Inquisitive learning	Human resources management and development/leadership
Acceptable risk welcomed	Leadership
Stretch goals	Process and quality improvement

Taken in summary, the attributes of structural fluidity represent observable expression of the Baldrige criteria. Note that the significant share of the responsibility for ensuring fluidity rests with organizational leadership. As we will note throughout this discussion of cultural characteristics and the human dimension, leadership is the absolutely fundamental ingredient in creating a high-performing corporate culture.

Measurement

High-performance organizations value measurement of performance. The dedication to measurement exhibited by these organizations bears close scrutiny. It is not a dedication to quantity measurement—measuring for the sake of measuring—but a devotion to gathering data and other information of importance to the organization. Many mediocre organizations can be noted to measure lots of things; it simply is a differentiating characteristic that high-performance organizations know not only how to measure but also what to measure. These organizations have discovered that progress is determined by more than time and balance sheet alone. More appropriately, they define their progress by assessing movement in the right direction. Once direction is known, the notion of how fast and how much can be quantified and thus measured (see Figure 7.3).

FIGURE 7.3. MEASUREMENT ALIGNED WITH EXTERNAL EXPECTATIONS.

Attributes	Dimensions
Direction focused	Process and quality improvement/leadership
Strategy inspired	Strategic planning
Values/ethics	Leadership
Financial and time data	Information management/process and quality improvement

That is not to say that financial or time-limited data are not important. To the contrary, high-performance organizations recognize the importance of financial outcomes, but they are not driven by a single-minded reverence for the balance sheet. Instead they are inspired by strategy, opportunity, and long-term vision. The jeopardy of judging progress by quarterly financial results is that emphasis is placed on short-term financial results as opposed to fostering a long-term commitment to the successful achievement of mission and vision. The fact that a singular focus on excellence in mission achievement promotes successful financial results is an interesting and somewhat ironic twist. High-performance organizations, though not driven by financial results, tend to derive better bottom lines than their less mission-focused colleagues (Chappell, 1993).

Concomitantly, these organizations have clearly identified values that are transmitted to all stakeholders. These values provide direction and guide organizational behavior. Typically, values support innovation, creativity, and, most important of all, integrity. Breaches of ethics, regardless of outcome or source, are not tolerated. It is important to remember that just as organizational heroes promulgate cultural values and give clues about organizationally sanctioned behavior, leadership's reactions to ethical breaches speak volumes about the actual intent and impact of such values.

Most of these organizations reflect an "eyes on the prize" mentality that is mission focused, not greed based. None of these organizations, whether for-profit or not-for-profit, is ignorant of the bottom line. To the contrary, high-performance organizations tend to reflect a healthy bottom line, but it is clear that this financial benefit is an outcome, not a goal. If business is done right and done well, a healthy bottom line will result as a measurable outcome of successful goal attainment.

It is important that the profit versus nonprofit status designation in high-performance organization be understood to be a tax status designation only.

All of these organizations, whether for-profit or not-for-profit, operate similarly. It is mission that varies from organization to organization.

Just as the bottom line is seen as an outcome and not the goal, time and revenue are treated as information sources but do not drive decision making. Certainly decision making involves consideration of the vital role of time and revenue, but decisions themselves are a reflection of a matrix of the organization's values, mission, and long-term goals. Measurement against strategies and long-term plans uses time and revenue as data-based indicators of progress.

High-performance organizations also measure their progress with attention to their core values—those that do not evolve with time and respond to the environment, but rather remain constant. High-performance organizations select a few key values and hold to them tightly. The way in which values are expressed is through the strategies they use to support value achievement. The kinds of activities used to achieve values and mission, however, may evolve.

For example, one of our clients is a major children's medical center in the Midwest. It was founded as an inpatient medical center with a heavy emphasis on research and medical education. Very clearly during the earlier incarnation, the organization's values were grounded in patient care provision in an inpatient hospital setting. But this high-performance organization has been able to adapt successfully to the mandates of the managed care environment's emphasis on outpatient and ambulatory care and has resolutely progressed toward delivery of healthcare resources and technology on an outpatient and distributive basis. Thus, the time-honored value for inpatient care has been supplanted by a more realistic value in terms of the current environment: a value for care where and how patients need it—in this case, an outpatient, ambulatory setting. But the core value of the organization—provision of high-quality healthcare for children—has remained constant. It is the second-tier values—the "how we do what we do"—that varies with the demands of the environment being served.

Leadership

Much has been said about the quality of leadership and its importance to organizational success. Unfortunately, much of the literature has painted a picture of the prominence of a charismatic leader. It holds that without a "big personality" at the top, the organization is doomed. Nothing could be further from reality.

The high-performance organizational grid evaluation developed by Compass Group, Inc., an organizational effectiveness consulting firm, reveals that leadership is a very real, almost tangible commodity at these organizations, but real leadership extends from the CEO's office through the senior levels of the organization, penetrating deep through the organization to the front line. Thus the term *leadership* does not denote position but rather connotes an array of behaviors appropriate to organizational success. The template allows an examination of these behaviors broken down into categories requisite to organizational success.

The Compass Group Leadership Performance Template is an aggregation of assessed attributes demonstrated by effective executives. It reflects more than a decade of organizational assessment. Although assessment techniques have been gradually refined, the common attributes of outstanding leaders have remained remarkably constant over time and across organizations.

In a high-performance organization, senior leadership sets vision and direction. Leadership is then able to extend vision, direction, and strategy to gain collective organizational commitment to mission fulfillment. This means that organizational leaders spend prodigious amounts of time and effort communicating to all constituents a few core themes requisite to organizational success. Through demonstrated behaviors and determined communication, leaders gain collected commitment to core values and vision. Consistency in communications rather than style or charisma seems to be the important indicator of this leadership attribute. Indeed, the presence of the big personality with flamboyant style and excellent presentation skills can be organizationally destructive if the leader is unable to hold a consistent theme and gain organizational commitment. Organizations are better served by less flamboyant leadership with a steadier commitment to direction and reinforcement of that direction. Additionally, organizations dependent on one big leader are at significant risk should that leader depart.

The leaders of high-performance organizations see their role as mobilizing the organization's energy toward vision attainment. This mobilization occurs through active delegation and empowerment of subordinates. High-performance organizational leaders understand that they cannot do it all themselves, and they do not want to. They work hard to energize others in a collective commitment to organizational vision. These leaders are able to delegate well and expect subordinates to fulfill their responsibilities.

Yet leaders also frequently exhibit a peculiar talent for maintaining what Peters and Waterman (1982) call "loose/tight controls." Leadership empowers people to pursue the organizational mission, and with a great deal of trust, by maintaining clear lines of accountability. People must be accountable and responsible for the decisions they make and outcomes that are, or are not, achieved. This may mean significant amounts of time watching and guiding details of a project at critical times, with easy and open communication between leadership and staff.

Leaders in high-performance organizations perceive quality of work life as a core responsibility. They set the tone for the organization and role-model behaviors supporting the desired work life. The quality of work life has a phenomenal impact on the ability of employees to attain the ultimate organizational vision. The energy drain, loss of focus, and malaise that result from employees' trying to function in an environment that is perceived as abusive, internally competitive, or highly stressful is phenomenal. Thus, even though these organizations are fast-track and can move with agility, work life issues can affect their ultimate performance. Work life issues are significant and must be attended to by the leadership with great respect and care (Ryan and Oestreich, 1991; Fenner and Fenner, 1989).

Figure 7.4 shows how leadership attributes align with external expectations.

Comfort with Paradox and Conflict

Much of what we have discussed to this point about high-performance organizations has reflected paradoxical balances. The ability to maintain "loose/tight" controls and the capacity for holding to unchanging values as direction and strategy are reshaped indicate that high-performance organizations are not only tolerant of paradox, but thrive on it.

Paradoxical balancing helps these organizations keep a focus on the bottom line while pursuing a vision of mission fulfillment. It permits these organizations to balance opposing forces such as "loose/tight" controls, the necessity for centralized authority accompanied by increased empowerment, the ability to build core competencies yet continually explore new technologies, the ability to support broad vision with tight strategy, and, most important, the ability to maintain both the short-term and the long-term focus. How do these organizations embrace these paradoxes? The answer is through the comfort level of their leadership, the leaders' ability to balance seemingly contradictory forces and to tolerate ambiguity (see Figure 7.5).

FIGURE 7.4.   LEADERSHIP ALIGNED WITH EXTERNAL EXPECTATIONS.

Attributes	Dimensions
Behavioral base	Leadership/human resources management and development
Vision and direction	Leadership/strategic planning
Energy and direction	Leadership/process and quality improvement
Key process focus	Process and quality improvement
Quality of work life	Client/patient rights and satisfaction

FIGURE 7.5. COMFORT WITH PARADOX ALIGNED WITH EXTERNAL EXPECTATIONS.

Attributes	Dimensions
Balanced paradox	Leadership/process and quality improvement
Conflict management	Leadership/client/patient rights and satisfaction
Dignity/respect	Leadership/client/patient rights and satisfaction

In addition to feeling comfortable with paradox, high-performance organizations are able to thrive on conflict. The reason lies in organizational values that distinguish between constructive conflict and destructive confrontation. High-performance organizations exhibit a conflict of ideas, a battle in the marketplace of knowledge, and the ability to support competing inquiry and test the applicability of inquiry in the real world simultaneously. The classic collegiate dictum, "We agree to disagree," seems to be an unspoken model in these organizations.

What is important to note is that conflict and disagreement are over ideas, strategy, research, and methodology. These organizations continue to exhibit a high degree of interpersonal respect. Thus, organizational conflict is in the marketplace of ideas, not between personalities.

Where conflict is encouraged, where conflict is discouraged, how it is reinforced, and how it is distinguished make the difference between high-performance organizations and other entities. Interminable warfare, political backstabbing, and just plain personal degradation are not tolerated in high-performance organizations, and they are extinguished when discovered. Rather, these organizations put great resources, time, and respect for learning into projects and then encourage several competing ideas to "duke it out" on the battleground of merit. It is not unusual to see seemingly opposing research and development activities occurring simultaneously. As the research gets closer and closer to maturity, proponents may be pitted against one another to demonstrate the relative merit of their activities. Yet never is this competition conducted through personal attacks, politics, or other organizationally destructive methods. There is a strong unspoken, yet continuously reinforced, value on individual dignity and personal respect in the high-performance organization. Healthy competition is welcomed, and destructive interpersonal conflict is reduced.

Deep Culture

High-performance organizations have an almost palpable corporate culture. There is thematic congruency about everything they do within their corporate culture. For example, Procter & Gamble of Cincinnati is said to exhibit the "P&G way." Procter & Gamble is a strong presence in the community through a continuous message about its culture and the way in which it reinforces that culture. In fact, staff at Procter & Gamble are frequently termed "Proctoids," and they are easily recognizable in the community as P&G employees. What makes for this deep culture? A closer look at P&G gives some strong clues.

First, the organization intensively attends to its values. As one international organization after another has been muddied with allegations of bribery and favor currying in financial dealing with host countries, P&G is conspicuous by its absence of mention. Procter & Gamble puts much time and attention into the research and development of its products and then brings a formidable marketing strength to promoting them. But nowhere in this extensive system of support for the launch of products is there any tolerance for wavering from a corporate value of integrity and honesty. P&G simply would not find itself in a position of having to defend a bribe given to an official to make way for its product in a marketplace. In fact, it has walked away from deals that did not pass the "smell test" when it comes to the integrity of corporate reputation.

Deep culture is more than the collective external behaviors of an organization; it is a system by which the organization imbues cultural values in its members. Thus, new recruits to high-performance organizations participate in extensive formal and informal systems for helping them see and accept, and later demonstrate, the culture of the corporation. This takes the form of extensive orientation, mentoring programs, proctorships, informal and formal social ceremonies, and celebrations that give messages to newcomers about the organization's culture. Deep culture is designed deliberately. Leadership and membership in the organization pay homage to and continually seek to reinforce the cues that reflect the organization's culture.

At the same time, behaviors that fly in the face of organizational values, that conflict with its culture, are quickly extinguished. Violations of pervasively held corporate values are dealt with fairly and rapidly. The true test of this attribute is when in the discovery of such an infraction there is also a concomitant cost to the company. The ability to swallow hard and pay the cost is a measure of leadership's ability to live the spoken values of the organization. Leaders embrace deep culture in an organization through "walking their talk."

Many organizations celebrate a propertied value of employees formally in their literature. It is not unusual to see organizations tout that their most important attribute is their employees. Yet these same organizations continually violate that value through promulgation and enforcement of demeaning personnel policies, tolerance of disrespectful and destructive interpersonal behavior, and implementation of an attitude of the disposable employee in difficult financial times. These organizations frequently try to cut their way to prosperity during difficult times by perceiving, and treating, their employees as an expense and not a resource.

For example, one organization that did not make the cut of high performance on several indexes has a strong internal and external publicity campaign related to organizational excellence. It purports to attain this excellence through the dedicated service of selfless employees. Yet this organization, a medical center, regularly uses downsizing and outplacement as a tool to meet short-term financial challenges. The conundrum here is impossible to balance for employees. The message contains too much cognitive dissonance: the spoken message is, "You are important to us," but the behaved message is, "Things get rough, we get tough, and you get tossed." Thus, the organization's culture is truly reflected in its behavior. The culture is focused on short-term financial goals regardless of messages sent to employees. As the literature now is beginning to demonstrate, too many corporations have tried to cut their way to financial success and have failed in the doing.

Deep culture is also reflected in the organization's symbols, structure and architecture, language, and overall ambiance. One organization that reflects successful attainments of high-performance status is a multiservice Catholic welfare agency. Its values are backed by its mission: belief in the strength of family, service to humanity, and a healing ministry. Its values are reflected in many ways, from the always respectful manner in which clients are greeted to the manner in which employees are oriented, with high degrees of respect and compassion.

In addition, all of the clues in the environment of services reflect the organization's commitment to its values, mission statement, Catholic heritage, and concern for human dignity. Organizational commitment is reinforced through displays of mission literature, religious artifacts, and other indicators of the organizational values: logos, identification cards, business cards with inspirational messages, and warm and heartfelt greetings in every interaction. You know that the organization is committed to its cultural roots in Catholic charity and human dignity because these roots are demonstrated in interactions with the members of the organization and communicated formally and informally throughout the organization.

The rule of policy and procedure in reinforcing or undermining the establishment of deep culture in organizations is frequently overlooked and merits attention. The way one manages policy and procedure speaks volumes about real organizational values. An example was provided by a medium-sized health system client that could not understand why an ambitious program to revitalize the organization had not been more successful. This organization highly touted its team-based management initiatives. And within the organization, many teams had been formed to pursue important organizational initiatives. There were new product teams, new market teams, and process reengineering teams galore. Even senior management labeled itself "the leadership team." Yet scrutiny of the organization's policies showed that it had a very competitive pay-for-performance compensation system. Compensation was based on the attainment of organizational goals through individual contribution to organizational productivity. Thus, the very teams the organization sought to support were undermined by the organization's own compensation system—a proverbial "two steps forward, one step back" to the provision of deep culture.

Organizational alignment—the matching of policy, procedure, behavior, goals, and reward systems to culture and vision—is key to building the deep culture of a truly effective organization (see Figure 7.6).

Stretch Goals

High-performance organizations have a very low tolerance for small vision. Achievement is conceptualized within a grand scheme of things. These organizations do not seek simply to survive. Their strategies, planning, and overall organizational efforts are dedicated to broad ambition and a sweeping vision. Marketing strategies may reflect an attempt for dominance in a market, segmentation from the mainstream or true niche status, or a "take no prisoners" market strategy for product launching. The notion of incremental progress is not alien to these organizations; it is just that incremental progress is assumed. Goal achievement is supported by everything the organization does.

High-performance organizations do not simply enter a new market without a vision for market position. Rather, they invade a market with the commitment to market dominance in the long haul. Vision for these organizations is directly attached to thriving in their chosen marketplace. Survival is simply not on their radar screen; it is given. That many organizations see themselves in a struggle for their very existence is unfortunate. High-performance organizations are refreshing in that they do not even note where they are surviving because their commitment is to thriving. In a time of small ambition, shrinking margins, and category creep for success, it is indicative for a high-performance organization to see strategy supporting market dominance or true market differentiation whether the organization is for profit or not for profit.

These organizations exhibit low tolerance for internal competition yet thirst for external competition, but without overfocusing on edging out the competition. They take a radical view of the industry and seek to provide breakthrough value

**FIGURE 7.6.   DEEP CULTURE ALIGNED
WITH EXTERNAL EXPECTATIONS.**

Attributes	Dimensions
Intensive values	Leadership
Walking the talk	Leadership
Stakeholders valued	Leadership/client/patient rights and satisfaction
Evidence of culture	Human resources management and development/process and quality improvement
Alignment	Leadership/process and quality improvement

for the customer. They are agile and swift regardless of size, and they are able to reposition and adapt to environmental changes, frequently before their competitors are even aware that the environment has changed. An excellent example from a high-performance organization has been the speed with which Procter & Gamble moved into tight correlation and inventory management for its client Wal-Mart when no one else had this strategy on their drawing boards. The nimble implementation of its inventory management system enabled Procter & Gamble to be a dominant provider of product to the nation's largest retail chain—not a bad position from which to compete.

One of our healthcare clients has as its vision to be not the largest provider of outpatient ambulatory services in a region but the only provider. Thus, all of its efforts are dedicated toward achieving total market dominance. All organizational strategy and resources are aimed at quickly aligning services to achieve this end.

The alignment of stretch goals with external expectations is shown in Figure 7.7.

Execution with Excellence

High-performance organizations value quality and create high expectations for attainment of quality. Formal and informal systems support quality achievement. Frequently formal improvement processes such as Total Quality Management are accompanied by informal, yet highly visible reinforcement of the quality expectations such as responsiveness to customers. Execution with excellence goes be-

FIGURE 7.7. STRETCH GOALS ALIGNED WITH EXTERNAL EXPECTATIONS.

Attributes	Dimensions
Broad vision and ambition	Strategic planning/leadership
Thriving	Process and quality improvement
Externally competitive	N/a
Leadership/information and analysis	Leadership/information management/process and quality improvement

yond evaluating and organizing to form a policy and procedure, and promulgation of performance measurement. If it cannot be measured, it did not happen (or is it, as one of us is fond of saying, "In God we trust; all others bring data"?).

High-performance organizations are obsessed with measuring quality and are committed to continuously improving. These organizations also value the nurturing of breakthrough strategic implementation (Gaucher and Coffey, 1997). They understand that what is dazzling excellence today is minimal entry into the marketplace tomorrow and stale the day after. Thus, they are focused not only on innovation but on continuous improvement of existing processes and products. This alignment of execution with external expectations is shown in Figure 7.8.

In fact, organizational hero myths are frequently related to attainment of exceptional excellence and moments of crisis in customer service that demonstrated the organization's commitment to quality. The success of these organizations is celebrated, but the celebration is brief, for these organizations are uncannily aware of how rapidly obsolescence takes hold if there is not continuous commitment to improvement, evolution, and excellence. Although time is not an absolute measure of performance, these organizations are acutely aware of the transitory nature of success. Remember how Humana took the healthcare marketplace by surprise and owned a large chain of hospitals before most others knew that proprietary healthcare was even a trend? Now note that Humana as a provider of acute care is a mere memory, having been absorbed by today's for-profit provider engine, Columbia/HCA.

The Attributes Summarized

High-performance organizations invest heavily in organizational learning, from very formal research and development and training programs to informal expectations that stakeholders maintain competency in their fields and in the marketplace. These

FIGURE 7.8. EXECUTION WITH EXCELLENCE ALIGNED WITH EXTERNAL EXPECTATIONS.

Attributes	Dimensions
Quality obsessed	Process and quality improvement
Measurement focused	Information management/process and quality improvement
Breakthrough strategy	Strategic planning/process and quality improvement
Obsolescence aware	Leadership/strategic planning

organizations cultivate an environment of learning where reading, listening, searching outside the organization, going on field trips, and generally sponging from the environment whatever competencies the organization can are rewarded and celebrated (Chawla and Renesch, 1995).

These organizations bear one additional attribute—one that is difficult to assess objectively and yet is almost palpable to participants within the organization: these organizations are fun. They are stimulating, creative, and always reinventing themselves. It is exciting to be affiliated with a high-performance culture, and one gets the sense that important things, regardless of business, are happening.

Applying the Model

Unlike more concrete measurement indexes—for example, the application of externally derived benchmarks against internal clinical outcome data—cultural and human dimension characteristics of high-performance organizations are much more difficult to see. Design, implementation, and evaluation of these characteristics require the ability to lead and manage with less than objective, empirically provable data as guides to progress. Yet even in this very soft area, there are ways to measure progress and outcomes, and this does not need to be as squishy and imprecise as "nailing Jell-O to the wall," as one of our clients termed the process. Model application is simply a matter of asking the right questions, in the right way, and with sufficient breadth, depth, and frequency to ensure reliable answers. The grid in Appendix B illustrates ways in which to assess these less than concrete attributes.

CHAPTER EIGHT

ANALYZING CORPORATE CULTURES

Continuing the analysis process begun in Chapters Four and Six, we now look at applying the criteria to the dimensions defined in Chapter Seven. Your attention while completing the analysis should be drawn to opportunities to assess the following areas:

- Gaps in performance from the ideal described as compared to the organization you are analyzing. These are often referred to as gap analyses.
- Duplications of functions, services, or attributes. Duplication of attributes is not inherently good or bad, but simply descriptive of the current state. Congruence of the criteria of bias for action, for example, may be a strength in a potential integration. Conversely, overreliance on the criteria reflected in teams and task forces could represent a potential weakness. Each requires the judgment and discretion of planners and leaders.
- Conflicts within the criteria. Potential affiliates with widely divergent cultures and implementation of human dimension criteria have a great deal of deliberative work to accomplish to ensure an effective integration.
- Appropriateness in relation to the idealized integrated organization being sought. Building a new culture affords an opportunity to design and implement systems using a fresh approach.

We suggest beginning with a high-level overview assessment and then proceeding to more detailed analyses. Look first for the major gaps, overlaps, and obvious issues that must be addressed. This first step may be conducted by a representative group of administrative and clinical leaders and staff who are familiar with data for the respective organizations. Since the overview analysis is often based more on perceptions than analyses of data, we suggest including people with different perspectives to avoid tunnel vision. The second stage is an intermediate level of detail and should be based more on data. The highly detailed analyses should be reserved for topics requiring detailed assessment.

Different types of people may be involved with the analyses of culture and the human dimension depending on the purpose and goals of the analysis—for example:

- Executive leadership and staff support for organizations considering new alliances and relationships
- Staff internal to a multifacility organization but outside the division being analyzed
- Joint participation of the organizations involved in the analyses
- Representatives of various stakeholder constituencies, such as physicians, management, employees, and corporate staff
- External reviewers, such as visiting professionals or consultants
- Official reviewers from the external accrediting, licensing, or certification organizations

The Analysis Tool

We continue to use a common format for the analysis tools throughout this book. Clearly, the tools should be tailored to the needs of your organization, and the purposes of the analyses. A sample tool for analyzing culture and the human dimensions is illustrated in Figure 8.1.

Selected key criteria are listed for each of the seven dimensions. At the bottom of each list is space to add other priority criteria for your organization. Also, note that there are some blank columns. These columns are used for the combined analysis matrix described in Chapter Ten.

Columns d, e, and f are illustrated for scores related to each of the important criteria. The score for your organization is entered in column d. Columns e and f are used for scores of partners or competitors, depending on which organizations are included in the analysis. Clearly, additional columns could be added to the spreadsheet for additional partners and competitors. Especially for the initial

FIGURE 8.1. ANALYSIS OF THE INTERNAL ORGANIZATIONAL CLIMATE.

Major Section		Important Criteria	Criteria Scores Score (0–4, 4 = high)			Comments
Number	Name	Description	a	Your Organization		
			b	c	d	e	f	

Scoring of criteria (fill in the boxed areas):
0 = No competence, coverage, or success, or not applicable
1 = Minimal competence, coverage, or success
2 = Partial competence, coverage, or success
3 = Significant competence, coverage, or success
4 = Substantial competence, coverage, or success

3	Internal Organizational Climate	**Structural Fluidity**
		Bias for action
		Turnover in point man
		Teams/task forces
		Customer focus
		Outcomes measurement
		Creativity valued
		Inquisitive learning
		Acceptable risk
		Stretch goals
		Other priority criteria, identified by organization

		Measurement
		Direction focused
		Strategy inspired
		Values/ethics
		Financial/time data
		Other priority criteria, identified by organization

FIGURE 8.1. ANALYSIS OF THE INTERNAL ORGANIZATIONAL CLIMATE. (continued)

Major Section		Important Criteria	Criteria Scores Score (0–4, 4 = high)					
Number	Name	Description			Your Organization			Comments
	a		b	c	d	e	f	

Leadership
Behavioral based
Vision and direction
Energy and direction
Key process focus
Quality of work life
Other priority criteria, identified
   by organization

Paradoxical
Balanced paradox
Conflict management
Dignity/respect
Other priority criteria, identified
   by organization

Deep Culture
Intensive values
Walking the talk
Stakeholder valued
Evidence of culture alignment

FIGURE 8.1. ANALYSIS OF THE INTERNAL ORGANIZATIONAL CLIMATE. (continued)

Major Section		Important Criteria	Criteria Scores Score (0–4, 4 = high)			Comments
Number	Name	Description	Your Organization			
	a		b	c	d	e	f	

Other priority criteria, identified
by organization

Stretch Goals
Broad vision and ambition
Thriving
Externally competitive
Leadership/information and analysis
Other priority criteria, identified
by organization

Execution with Excellence
Quality obsessed
Measurement focused
Breakthrough strategy
Obsolescence aware
Other priority criteria, identified
by organization

general analyses, a simple scoring system is recommended. The question you are trying to answer is how well your organization, partners, or competitors are addressing the different components of the taxonomy. The following five-level scoring system is illustrated in Figure 8.1:

0   No competence, coverage, or success, or not applicable

1   Minimal competence, coverage, or success

2   Partial competence, coverage, or success

3   Significant competence, coverage, or success

4   Substantial competence, coverage, or success

A column is included for comments. Although any comments may be helpful, the entries in this column are most useful if they address the need clarification and the need for more detailed analyses, and in particular the type of comparative information that is desired.

The Analyses

This section provides example interpretations related to culture and the human dimension. A key purpose of these analyses is to expand your thinking beyond your current services or even the services provided by your partners and competitors.

Structural Fluidity

Nine criteria are listed in Figure 8.1: bias for action, turnover in point man position, teams and task forces, customer focus, outcomes measurement, creativity valued, inquisitive learning, acceptable risk, and stretch goals. Clearly, each of these can be addressed at much greater or lesser levels of detail. For the general overview analysis, the following questions can be used to assign a score:

- Does the organization formally address and make an impact on this topic? For example, are there recognized and supported teams and task forces?
- What data have you collected and analyzed on the topic? For example, most healthcare organizations have virtually no data on or understanding of the degree of acceptable risk encouraged for participants. But many effective organizations collect and provide detailed analyses of outcomes measurement.
- To what degree has the organization demonstrated success in making an impact on the topic?

- Are there inappropriate duplications or conflicts? For example, some divisions or departments may be highly rigid, centralized, and authoritarian and might be scored 0 or 1 on the acceptable risk criterion, while others may be very risk enriched and teamwork oriented and might be scored 3 or 4 on the same criterion. An overall organizational score could reflect a blended score, or might use the gap or differences between areas as an opportunity to address internal inconsistencies.

In the comments section, it may be helpful to identify the major gaps, duplications, conflicts, and issues, so they can be addressed.

Measurement

The four criteria illustrated in Figure 8.1 are direction focused, strategy inspired, values/ethics, and financial/time data. For the general overview analysis, the following questions can be used to assign a criterion score:

- Are all criteria reflected in the organization?
- Are there inappropriate duplications or conflicts?

In the comments section, you may want to note the gaps of topics not addressed, duplications, conflicts, and issues.

Leadership

Five criteria for leadership are illustrated in Figure 8.1: behavioral based, vision and direction, energy and direction, key process focus, and quality of work life. The score for each criteria can be based on questions like the following:

- Do leaders demonstrate and exemplify this behavior?
- Has the organization dedicated resources to assessing any of these criteria?
- Are there inappropriate duplications or conflicts, gaps, or redundancies?

Evaluation of organizational leadership can be a delicate, risky process for subordinates. For most accurate results, this assessment may be best completed by either the leadership team or external consultants. Each organization must make an individual determination as to best approach and strategy. Do note that ideally leadership is not solely the prerogative of top management but extends throughout the system. In the comments section, you may want to note specific areas for follow-up.

Paradoxical

Three criteria for assessing comfort with paradox are illustrated in Figure 8.1: balanced paradox, conflict management, and dignity/respect. The score for each criterion can be based on questions like the following:

- Does the organization exhibit this attribute?
- Can you note examples or illustrations of this property in process?
- Are there inappropriate duplications or conflicts?

The comments section may be used to identify topics and issues for follow-up.

Deep Culture

Four attributes are illustrated in Figure 8.1: intensive values, walking the talk, stakeholder valued, and evidence of culture alignment. The score for each criterion can be based on questions like the following:

- How well does the organization reflect this attribute? Are there environmental cues that reinforce core values? Does behavior reflect stated values, or are there significant gaps between rhetoric and reality?
- Are there inappropriate duplications or conflicts? The act of merging organizations creates huge conflicts among these basic behaviors and values.
- Do staff understand and believe that spoken policy conforms to behavior?
- Are there complaints about the process? If so, the score will be lower.
- What data exist to demonstrate that the process is successful?

The comments section may be used to identify further information.

Stretch Goals

This dimension is the one most likely to require modification by your organization. There are four key criteria listed in Figure 8.1 to assess: broad vision and ambition, thriving, externally competitive, and leadership/information and analysis. The score for each criterion can be based on questions like the following:

- Is vision a known and respected driver of decisions?
- Is there urgency to exceed industry standards and success measures?
- Do you collect and analyze competitive data and benchmarks?

- Are leaders consumed with pushing the organization to attain larger and more expansive achievements?
- Have you received industry recognition for extraordinary achievements?
- Are there inappropriate duplications or conflicts?

Execution with Excellence

There are four criteria listed in Figure 8.1 that are used to analyze this dimension: quality obsessed, measurement focused, breakthrough strategy, and obsolescence aware. The sample questions in addressing these criteria may include the following:

- Has the organization received external recognition for excellence achieved?
- Are there consistently measured outcomes and processes, assessed over time and across the organization?
- Can you demonstrate significant innovation or invention developed and implemented by the organization?
- Are there continuous efforts to renew and reinvent the organization?

Steps in Using the Analysis Tool

Now that you understand the approach to conducting the analysis and the use of the tool, your organization should tailor the process and tool to meet your requirements:

1. Decide the purpose of the analysis.

    Are you analyzing a single organization?

    Are you analyzing an existing healthcare system?

    Are you analyzing potential partners in a healthcare system?

    Are you comparing against competitors, or system partners, or both?
2. Determine the level of detail for your current assessment.
3. Review the criteria used in Figure 8.1 for each dimension. Do you want to add, delete, or amend any? Remember that using a higher level of detail is fine, but deleting a component may cause you to fail to meet the intended completeness of the analysis using the taxonomy.
4. Revise the analysis tool to meet your requirements.
5. Select the people to participate in the analysis.
6. Review the analysis tool, the operational definitions of terms, and the scoring method with the people involved with the analysis.

The completed analysis will serve as a useful tool in guiding integration readiness, appropriateness of potential partners, and opportunities for internal improvement in anticipation of readiness for integration. The integration assessment, as previously described on page 135, is illustrated in Figure 8.2.

Conducting More Detailed Analyses

The overview analysis typically is based primarily on perceptions and readily available data. Minimal effort is spent collecting data for the overview analysis. The risk, of course, is that the perceptions may be incorrect, so more detailed and quantitatively based analyses may be desired concerning selected topics. The same analysis tool, Figure 8.1, can be used. The difference is that the scores entered for each criterion are now based on more quantitative data. Appendix B gives sample questions to assess a finer level of detail in the internal organizational climate dimension.

Taking Actions Based on Analyses

The culture and human dimension analyses, conducted using Figure 8.1, will lead to a number of actions. The following examples are of the types of actions and additional analyses that your organization may undertake:

FIGURE 8.2. HORIZONTAL AND VERTICAL INTEGRATION OF THE INTERNAL ORGANIZATIONAL CLIMATE.

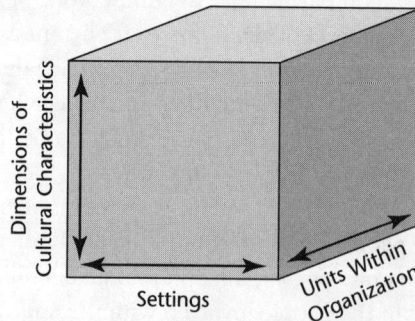

- Identify opportunities. By the time you have finished the analyses, you will almost certainly identify some opportunities for new programs, changed foci, additional development, or different ways of doing business.
- Identify duplications. You need to identify duplications within your system or among your organization and potential partners. Some of these duplications will be unnecessary and costly; others will be appropriate.
- Determine the appropriateness of the current operational culture. Appropriateness is difficult to evaluate because different people have different perceptions of appropriateness. It is incumbent on the leadership of the organization to determine culture as a critical element of organizational success.
- Identify issues and conflicts for resolution.
- Assess the compatibility of potential partners.
- Identify and prioritize areas for more detailed analyses.
- Identify approaches to fill gaps.
- Identify approaches to eliminate inappropriate duplication.
- Develop collaborations to maximize quality and cost-effectiveness.

The results of the analyses related to culture and the human dimension are one of the key inputs to assessing and developing an integrated health system. Failure to attend effectively to these vital forces can greatly retard or destroy potential partnerships, individual organizational excellence, and achievement. The scores for the specific criteria included in Figure 8.1 may be used in two different ways. If you chose to do the combined analyses at a more general level, then the specific results from the analyses discussed in this chapter serve as the basis for the scores assigned in the combined analysis table. Alternatively, the more specific criteria in Figure 8.1 may be incorporated into a more detailed spreadsheet replacing Figure 10.1.

CHAPTER NINE

SPEED BUMPS ON THE ROAD TO INTEGRATION

The process of system integration generally yields several surprises. Some of them are welcome; many of them are unwelcome, but most of these are predictable, at least to an experienced external observer. This chapter looks at the most frequent of these challenges—we call them speed bumps—on the road to integration, analyzes their origin, and proposes strategies to prevent or remedy their occurrence.

Every edition of *Modern Healthcare* seems to feature stories about failed mergers and acquisitions among hospitals and other healthcare corporations. Social services agency consolidation and integration attempts are no less vulnerable to missteps and failures. For each of these very public failures, there are numerous flawed but still limping integration efforts that nevertheless function but are less than optimal in meeting the intended goals for organizational optimization.

Analyses of these missed opportunities reveal the failed processes of these easily preventable problems. Remembering that hindsight is often 20/20 vision, we have taken this opportunity to apply a simple assessment matrix to less than optimal situations. We have also compared and contrasted these cases with standout organizational examples—the emerging systems that have captured the opportunities offered and avoided the pitfalls on the way. The assessment matrix, an additional application of the VIHS taxonomy (Chapter Three) and internal organizational climate (Chapter Seven), reveals integration speed bumps that can be arrayed into six categories of potential failure origin.

If success at integration is so important and failure's origin so easily discerned, why do so many bright and well-intentioned people find themselves in the midst of organizational crisis during integration? Perhaps Andrei Codrescu's thought, "Perspective is at the source of all knowledge. What we see depends on where we stand" (1996, p. 9), applied here will serve us well. It seems that once organizations begin the merger and integration process, much of the necessary objectivity is lost. It is as if once embarking on this venture, all ability to have perspective is neutralized. Thus, the benefit of analyzing others' experiences is critical to preparing for, or correcting, one's own efforts.

Case analysis of successful organizational mergers and acquisitions, and of less than optimal merger and acquisition activities in the field, has yielded a predictable typology of risk areas that merit attention as loci of integration challenges and pitfalls. We will review these risk points (speed bumps), discuss preventive strategies, and examine implementation alternatives that can prevent the painful outcomes of poor integration.

To achieve the cost efficiencies and increased effectiveness possible through service integration, providers must create a seamless delivery system: one in which the consumer of service does not experience any interruptions or diminution in service quality, effectiveness, or efficiency in the process of obtaining care within and across system components. Speed bumps become evident to the customer when service integration compromises effectiveness and efficiency. But speed bumps can exist and detract from integration effectiveness without customer awareness. Integration should not only result in seamless service to the customer but also should yield organizational efficiencies, operational economies, market share increase, and other strategic attributes. Speed bumps encountered frequently are evident in reduced efficiencies and effectiveness in customer service but may also be expressed in overlooking the advantages originally perceived as the rationale for integration. Integration for integration's sake and growing bigger just for the sake of being bigger yield little benefit. The goal of integration should be more than to become a behemoth. Thus, we will focus on the eight frequent areas of integration where speed bumps occur, analyzing how they occur, why they occur, and preventive strategies for each.

The Speed Bumps Defined

Let us look at each of the eight speed bumps in turn.

Executive Leadership

The most frequent failures in organizational integration are due not to failures of will, financing, or planning but to the inability to make an appropriate selection of executive leadership for the new, emergent organization. Governing boards of nonprofit organizations encounter a merger activity unprepared for the hardest decisions of all: who will be the senior executive and who will be on the senior executive team. Past loyalties, current pressures, and, most frequently, the inability to confront the communication of this news to the executives involved frequently delay the making of this critical decision. It is amazing how difficult it is for assembled boards to confront the task of succession. The determination of the executive leadership of the newly formed organization is the most important task of the organization's governing board. Yet it is the decision most frequently delayed, deferred, and fumbled.

Board Composition

Second only to the inability to determine executive leadership is the reluctance with which boards examine their own structure, membership selection, responsibilities, and governance prototype. The merger of boards of trustees is not unlike a marriage; there are family members who will fit well and have an important role going forward, family members whose affinity and goodwill is sought but who are not key to the continuing functioning of the new unit, and family members no one wants to invite to dinner because they significantly detract from the quality of the experience. Yet given the opportunity to structure a new, more effective and efficient board, it is amazing how frequently this issue of member selection and board structure and governance is deferred and delayed. Again this can be contributed to an inability to confront and communicate bad news. And again we have a significant deal-breaking impact.

Resources

The allocation of resources, determination of percentage of participation, risk ratios, audit accountability, and, most important, sources of external support confront all new integrations. Although significant effort is usually expended in the execution of due diligence examination, discussion about future resource acquisition and distribution is typically deferred until later in the game. Yet it is in the answer to the question, "How are we going to sustain and grow the system?" that the success of the new organization is determined.

Legal and Regulatory

The approval of a corporate structure, accreditation, and license strategy needs to be explored early in any discussion of integration of separate corporate entities. This is of particular import in instances where for-profit organizations are acquiring or merging with not-for-profit organizations. State officials, local community agencies, and regulatory agencies are looking with increasing disdain on the acquisition of not-for-profit resources by for-profit organizations and are asking very legitimate questions about the disposition of community charitable assets by for-profit agencies. These issues need to be fully explored in the initial stages of evaluation of alternatives.

Community

Particularly with long-standing, not-for-profit community agencies, the issues of community support, public relations, communications, and potential fundraising impact need to be explored fully. Communities that have spent decades building health and social services resources do not lightly part with the legacy they feel themselves to have created. The handling of the public side of any integration is resplendent with opportunities for failure. Frequently what occurs is reminiscent of the old joke, "The surgery was a success though the patient died." As integration is considered, it is best to ask, "What good is it to capture the efficiency and effectiveness of a merger or acquisition if in the process the most valuable coinage, public goodwill, is wasted?"

Mission and Vision

The historical commitment to mission frequently gets in the way of an organization's thriving in the future. The new organization needs to have a new, appropriate mission and vision to guide its future. This does not mean that the revered missions and legacies of the previous organizations are lost or abandoned. To the contrary, the new mission frequently builds on the legacy of past achievements and traditions. In fact there are certain "drop-dead" or deal-breaking issues that are confronted during merger and integration of organizations relative to existing mission and values. Corporate culture, religious and community ties, sanctions and restrictions from incorporation covenants, and the history and habits of organizations greatly affect their ability to form part of a new, future organization. Futures planning processes for integration that do not carefully examine the historical legacies of all participant organizations set themselves up for failure.

Process Failures

The design and implementation of a new organization, whether from the merger of two or more existing organizations or from acquisition of another organization, is a deliberative, delicate process. Missteps, leaked information, assumption of inaccurate information, or failure to complete the process diligently yet efficiently can doom or seriously flaw the successful outcome.

Physicians and Integration

Although physicians are not the only constituency affected by system integration, they are the most important one (beyond the patient or client of service). Certainly failure to attend to their pivotal role can spell doom to any effort, no matter how noble or innovative.

Executive Leadership

The board of trustees of any for-profit or not-for-profit organization has no greater responsibility than the selection, guidance, and evaluation of the chief executive officer. This is the first and foremost task of the board, yet the one that seems to be the most difficult to confront. When existing organizations approach the table to discuss integration into a new entity, they typically already have an entrenched chief executive officer (CEO). Each board may have deep affinity and loyalty to their current chief executive. This informal, historical, and frequently strong social bond greatly complicates the ability of organizations to designate a new CEO. Many mergers have collapsed on the issue of who is going to lead. The issue of CEO can be, and frequently is, the deal breaker in a merger. Acquisitions, of course, afford the opportunity for the acquiring organization's CEO to know from the beginning of the process that he or she will be the emergent CEO or will be appointing the CEO in a system-based acquisition. This is not true of mergers and integrations where the partners often have similar size, stature, and clout going into initial discussions. On hindsight, this seems to be an important yet surmountable detail. But several very public merger efforts have been derailed by the inability to resolve the issue of "our CEO against your CEO."

There are several strategies that can be used to avoid this pitfall. First and foremost is the leadership of the current executive officers in recognizing the problem and discussing its resolution prior to the initiation of further integration efforts. However, this rarely happens since the most frequent scenario is one in which each of the CEOs perceives himself or herself to be the more qualified for the new executive position. One of the larger system CEOs has a standing merger and acquisition policy that the incumbent CEOs cannot serve in that capacity unless it is a requirement of the acquisition. Her philosophy, borne out in results, is that generous severance packages and new starts elsewhere (sometimes within the same systems to capture evident talent) provide a fresh start for the emerging entity's integration into the system.

The CEO selection dilemma is also avoided when the initiation of the momentum and the key leadership for merger and system formation originate clearly with one organization's leader. Frequently one executive has both the vision and the risk-taking entrepreneurial ability to see and pursue the possibilities and potentials of a merger or integration. It is often this executive who naturally emerges as the leader of the new organization. There are, of course, challenges with the emergence of one executive from the affiliating entities. Stakeholders have concerns about objectivity, bias, history, or preconceived assumptions of style. Yet there are also significant benefits to this event, not the least being the clear message of reward for appropriate risk taking and venturing—skills that should serve the new, integrated organization well.

The following process represents an ideal design for selection of the CEO and executive team to lead the emergent organization. Like all other ideal designs, the pragmatic application will result in some alteration or modification, but it is a useful speed bump avoidance device to begin with an ideal set to model.

The leadership of the organization, composed of members of the board of trustees, current executives, and representatives of key organizational functions and constituents (consumers, payers, physicians), define, assign weight, and rank the leadership competencies critical to the organization's future success. This group must determine the key objectives for the organization's formation and the executive abilities imperative to objective accomplishment. In short, why is this organization in existence, where does it need to go, and what competencies must its leader have to get it there?

A normed, standardized selection tool such as Career Architect by Lominger Limited, Inc., is essential to the objective, efficient design of the competency template for executive selection:

1. This competency template is used as the base for the selection process. The search committee or its designated agent (consultant or executive search firm) uses the competency template as a guide to recruitment advertising, initial culling of potential candidates, and ranking of finalists by fit with the weighted competencies.
2. Selection of the chief executive occurs after assurance that finalists meet the expectations of the board leadership, not in accordance with past or current

loyalties to incumbent executives, avoiding one of the typical speed bumps that derails successful organizational integration.

3. Selection of the CEO is based on fit between the competencies as ideally designed and the assessed attributes of candidates for the position. A weighted ranking of competencies of candidates compared against the ideal template lends a high degree of objectivity and assurance of success to the selection process. The new CEO is selected in accordance with the competency template design.

4. Compensation systems for the executive reflect and reinforce implementation of the competencies as designed in the initial search process. The issue of compensation as an integral part of the initial executive selection process is too often ignored or relegated to the negotiation phase of final selection. Compensation parameters should be determined well prior to final candidate selection.

Board Composition

Decisions about board composition are frequent speed bumps experienced on this road. This is most unfortunate since the quality of board leadership and the assemblage of the requisite skills for this leadership are second only to CEO selection in import to the overall success of the emergent organization. The most frequently encountered impediment to this accomplishment is the lack of will to confront the difficult selection process. The inability to be forthright about the desirable attributes and skills of the new board is typically matched by a lack of courage in the communication of the perceived bad news of not being selected to the members of the previous institutional boards.

Yet there is an effective approach to board structure, design, and composition. First is the necessity of depersonalizing the process. It is imperative that all past, current, and future board members understand that their first and foremost responsibility is the accomplishment of the mission of the new organization. Different skills, competencies, and constituency representation assist in this accomplishment at different times. Each incumbent member of the previous boards must be able to demonstrate his or her comprehension of the importance of this "clean slate" philosophy for going forward by being theoretically willing to resign from the board as a precursor to formation of the new board. In reality, however, there will be planned membership overlap for continuity between the previous boards and the new governing body.

The new board of the emergent organization must be constituted according to its mission, not as a reflection of the two or more previous organizations. Too frequently the new board is seen as a representation of the previous boards. More disastrous is simply combining the existing boards into a new, large board for the new entity. This results in lost opportunity—the opportunity of retaining and recruiting the board members most appropriate to the new mission and challenges. It also frequently yields a board that is much too large to be effective.

Certainly retaining the history and experience amassed in the previous boards is important, but it is more important that the emerging organization have the board leadership required for success. The vital question is, What skills are needed to realize the new vision?

The ideal process for composing the emergent board is to structure a board design committee composed of representatives of the previous organizations' boards, executive leadership, legal counsel, and outside consultative assistance from a dispassionate, objective resource. Starting with a clean-slate philosophy, the process begins with the combined leadership of the organizations (board and executive) specifying the desired outcomes of the proposed merger or integration. "Begin with the ends in mind," a mainstay guide to any planning, is a requirement for the effective design of the new organization. Key questions to be asked include the following:

- What markets are sought? Who are the key payers? Who will be key to the future?
- What efficiencies are projected? Consolidations sought?
- What new services and settings will be proposed? What board knowledge will be required?
- Who are our vital external customers? Internal customers?
- What organizational competencies are fundamental to success?
- Who are our key providers of service (physicians and others)?
- What governance requirements for representation must be met?
- What will be the community benefit of integration?

From the responses to these and other key questions, the design group is able to draft a list of competencies sought for board members in the new organization. These competencies may include the ability to speak for and to major customers, such as leadership of local businesses, members of the surrounding communities, legal or financial leadership, consumers of service reflecting the diversity of the population, and governmental representatives. Physician and other provider leadership is also a vital constituency to include in consideration of board composition. Hospital executives have frequently sought to limit or exclude physicians as governance leaders, a logistical mistake since past, current, and future operational success is clearly dependent on physician leadership, support, and participation as key stakeholders in any healthcare system.

Beyond the issue of appropriate representation from sought constituencies, it is important to consider the personal attributes each board member must possess to contribute to the accomplishment of the organization's mission. Creativity, communication skills, discretion, analytical skills, knowledge of the marketplace, and other competencies should be important determinants of eligibility for board membership. Leaders should avoid the trap of appointing board members simply because they are nice, well liked, or economically well positioned in the community. The task of determining board composition is vital to organizational survival and should not be relegated to a political process or considerations of "payback" for favors granted during consolidation or merger discussions, though these practical and pragmatic considerations do impinge on the process.

Resources

The process of due diligence is one of the most effectively executed functions for the majority of consolidation efforts, yet the topic of resources planning requires much more than the careful exploration and disclosure of financial assets and liabilities. The fundamental decisions of planning for future-focused resource allocation will greatly influence organizational success. Certainly it is important to know what assets are available and what obligations exist, but it is more important to plan and target allocation of available resources to strategic future initiatives, as opposed to funding blindly what has always been funded or, worse yet, funding according to the influence and lobbying of organizational constituencies.

A resources allocation strategy encompasses more than the institutions' financial resources. In fact, the most undervalued yet critical resource requisite to organizational success is the energy and enthusiasm of human capital. Allocation strategies need to interweave the appreciation and utilization of human capital tightly in as diligent a fashion as is the expenditure of financial resources.

Resources allocation planning must be tightly keyed to organizational strategic planning. The decisions of how much is devoted to what are best based on knowledge of the fundamental requirements for success. These critical success factors are revealed through the strategic planning process. Again the mantra, "Begin with the end in mind," is applicable.

The strategic planning process can be simply summarized as follows:

• *Mission, vision, and values.* The organization's purpose for existing, its ideal destination, and the behaviors and standards it sets for its members are key determinants of the future. This is a critical first step requirement that provides the basis for all that follows. Four fundamental questions need to be asked: Where are we now? Where do we want to go? How do we get there? and How do we measure achievement? Mission, vision, and values determination, undertaken by the organization's leaders and communicated to all organizational constituencies, does not require time-consuming contemplation or laborious effort. But the process of committing to a guiding mission must be accomplished as a precursor to the details of strategic planning.

- *Goals and objectives.* A measurement system and accountability process are integral to accomplishing organizational mission. This step in the process answers the key question, How will we know if we're driving in the right direction? The setting of objectives to support accomplishment of broader goals and then the implementation of systems to measure and monitor objective achievement are critical. Gouillart and Kelly (1995, pp. 70–79) outline an excellent process for building a balanced scorecard as a measurement system.

- *Allocation of resources.* Resources allocation, control, and monitoring are based on the strategic imperatives designed in the planning process. Scarce resources are devoted to activities vital to success; new initiatives and old but less critical programs must compete for support against an evaluative template based on the critical success factors for the organization's future—not relative to any political process or continuation of tradition. "We've always done it" does not provide a rational reason for continuing to do it, but rather is one of the key phrases indicating potential failure during organizational integration.

Legal and Regulatory

The structuring of the legal entity for corporate sponsorship of the emergent organization is a decision best undertaken only after specialized legal counsel has been consulted. Too often organizations rely on the advice of general counsel—representatives who may have quite competently met the organization's past legal needs but who may not be best qualified for rendering this highly specialized advice. This error can result in disastrous consequences, both legally and operationally. Incorporation formats and structures are varied and complex. A few alternatives for consideration include for-profit, not-for-profit, foundation, private, and public structures. Each has restrictions and advantages, all require careful thought and consideration, and none should be lightly assumed. Different components of the VIHS may require different legal structures. Rapid-fire decision making at this point can result in months, or even years, of litigation and regulatory scrutiny.

Similarly, organizational merger and acquisition activity should not ignore the impact of change on accreditation and professional association membership.

The JCAHO, the COA, and various state departments of public health all require involvement and may mandate recertification or accreditation accompanying major organizational change. Careful communication with these and related agencies is imperative if time- and resource-consuming reviews are to be avoided.

Community

The coinage of goodwill accumulated by the majority of community-based health and social services organizations cannot be overvalued. Decades of volunteer effort, fund development, public relations implementation, and just plain good service to community members have resulted in the amassing of significant, positive, and frequently possessive feelings of community members toward "their" hospitals, healthcare facilities, and social services entities. This energy of community commitment is a real and valuable resource, both during difficult operational challenges and for creating momentum toward growth. Yet here again we frequently see opportunity lost and speed bumps created through insufficient sensitivity to the psychic investment of community members in the institution.

The first principle to address when considering community impact is that of no surprises. Absolutely nothing erodes community trust in an institution more than discovering that the organization's leaders are attempting to negotiate radical change in ownership or sponsorship without consulting with the community at large. And nothing sells more local newspapers than the cover story by an ambitious reporter who has gotten wind of ongoing negotiations to "sell our hospital."

This admonition does not mean that all considerations of merger or acquisition activity need to be managed in the public eye or only after first obtaining public comment. In fact, quite the contrary is true. The delicate initial phase of contemplation of any radical organizational change must be managed with the utmost discretion and attention to confidentiality. But when strategic decisions to pursue a change in direction are clear to the organization's senior leadership, a formal plan for managing community relations and communication is not only prudent but absolutely vital. Failure to pay serious attention to the real ownership that constituents feel in community organizations incurs risk for loss of loyalty, a commodity not easily rebuilt.

Additional public relations and legal issues can be raised if the delicate question of appropriate stewardship of endowment, public, and charitable assets is not managed effectively. Some hospital acquisitions have garnered the attention of the state attorney general, who is concerned that assets of community not-for-profit hospitals were being transferred to for-profit corporations (Japsen, 1996, pp. 4–5).

Again, competent legal advice is requisite to success, as is careful attention to management of community and public relations. Failure to attend to maintaining community support during and after merger or alliance activity can erode market share, deplete fundraising results, and offer strategic entry opportunities to competitors.

Mission and Vision

Failure to comprehend and respect the distinct institutional missions and histories of potential partners or participants in a new organization can doom the venture before its start. Classic examples of failure to respect origins and values abound. The standout example from our experience was the merger of a regionally prominent Catholic system with a similarly sized health system with nondenominational, community-based sponsorship. The requisite due diligence was complete, all public and internal constituencies appropriately consulted, and the formal roll out of a new name and identity were set to go when, in the course of operational integration discussions, the fertility clinic owned by the community system was reviewed. Suddenly the Catholic-sponsored system core value of sanctity of life was perceived as being violated—a deal-breaking threat. The existence of a small clinical service with very little volume or impact on the core businesses almost completely derailed the merger effort midstream. Only quick thinking—the fertility clinic was spun off as a separate corporate entity—managed to restart the process.

Institutional history and corporate culture are as important to the success of any venture in merger, acquisition, or affiliation, as are respect for and acknowledgment of, mission. An organization with a strong centralized tradition of top-down decision making and a command-and-control leadership style will have great difficulty effectively integrating with a more open, collaboratively managed institution. Assessment of differences in culture, style, and even terminology between potential partners and formulation of strategy to address these differences is imperative if success is to be achieved.

Process Failures

Failure and less than optimal outcomes are still frequent, even after completing all of the content segments correctly. Integration process problems plague effective achievement of the goals that originally motivated the activity. Process problems are defined as managing the flow of the activity, as opposed to the content of the intended result. The goal, of course, is to combine two or more organizations into a new, more effective, and efficient entity. The component parts of the content defined through the leadership and governance structure are relatively concrete when compared to the diplomatic task of managing the process of integration.

Process speed bumps include disclosure of planning discussions prematurely (recall the eager newspaper reporter), attempting to seek major financial results through indiscriminate downsizing without sufficient operational experience to evaluate key personnel, neglecting to seek input from or respect strongly held views of vital customers or constituencies, focusing on internal operations without attention to the new and developing demands of the marketplace, and similar lapses in leadership. Perhaps the most frequently encountered and damaging process problem of integration is failure to address the removal of organizational silos, or the creation of new, additional silos with the integration effort.

Healthcare, health, and social services systems frequently are divisible into silos: formal or informal organizational structures reflecting communication and affinity patterns for members of the organization. Silos reflect organizational departments, divisions, and related administrative structures that give form and discreet structure to complex systems. As with professional allegiance, department or other administrative silos serve to provide a constituency for those included, a manner of communicating with colleagues, and a method for reinforcing identity and functionality within the silo structure.

Organizational silos can create inefficient and ineffective divisions in organizations, also usually based on departmental affiliation or professional identity (Marszlek-Gaucher and Coffey, 1990, pp. 97–98; Gaucher and Coffey, 1993, pp. 153–154). For example, in most hospitals, physicians form a tight silo of professional identity, communicating, associating, and allying with physician colleagues at a far greater level than they might even with staff with whom they spend much more time and have more interaction. Similarly, nursing staff form a tight identity group, excluding nonnursing personnel regardless of preparation.

The compartmentalization of organizations into administrative structures renders efficiency for the purposes of traditional managerial functioning, yielding a means of maintaining control, audit, and evaluation functions. However, there is also a very real adverse impact of silos in organizations through the creation of frequently insurmountable barriers. These barriers—the speed bumps erected by departmental or professional communication and work patterns—impede cross-functionality. Staff become more interested in the needs, goals, and loyalties of their respective group, department, or division than in the needs and objectives of the entire organization or the goals and requirements of the organization's customers. Quickly we find that communication flows up and down within the silo but not across silo barriers. Vital information does not follow the functions required to serve customers but instead follows the geography of the silo.

Customer focus is replaced by a focus on department, division, or professional affiliation. The professed purpose of the organization—customer service—takes a secondary priority to the goal of maintaining departmental integrity or professional autonomy. Frequently this subversion of organizational purpose is both unintentional and unacknowledged. It is a by-product and outcome of the professional and organizational strategies inherent in hospitals, health, and social services systems. The traditional hierarchical structure has permitted institutionalization of management control and efficient allocation and analysis of resources. The lost focus on the primacy of customer service and the need to maintain cross-functional systems has been unfortunately relegated to a second-tier status.

The seemingly efficient and effective patterns of traditional management structure directly impede the actual primary organizational goal. The impact of silos can be readily observed during organizational assessment, the activity of analyzing organizational effectiveness. Frequently observation of deteriorating and dysfunctional organizations reveals that internal silos replace customers as centers of attention for participants. In fact, organizations in severe distress will display competitive behavior between silos in vain attempts to maintain viability in the face of decline, all the while ignoring the centrality of the customer to survival and success.

Organizational silos at work in healthcare can be depicted by tracing the progress of particular patients, the actual customers, through the care delivery systems of hospitals. Although the patient is progressing through the system in a linear fashion, consuming the services of a variety of organizational departments according to diagnosis and need, the structure of the organization maintains control and communication within the silos, as is traditional with managerial functioning. The patient then encounters a variety of experiences with variances in quality, timeliness, and competence according to the silo within which service is delivered. Most organizational disruptions, disconnects, and problems occur at the "hand offs," the point at which service to the patient is transferred from one department, or silo, to another. Whether it is the lack of continuity in communication, missed opportunity, or interdepartmental rivalry, the lapses in care continuity are more prevalent at the edges of silos. Figure 9.1 illustrates the dysfunctional impact of organizational silos on customer service.

Note that communication, identity, and affinity flow within the silo, not across the organization, thus increasing the possibility that the patient or client will experience a disruption in service. For example, physicians are more likely to identify and communicate with their colleague physicians (within the professional affiliation silo) than to cross silo boundaries and communicate with members of another organizational silo. Similarly communication flows more easily within the silo embodied in the laboratory department than it will across silo boundaries to

FIGURE 9.1. ORGANIZATIONAL SILOS.

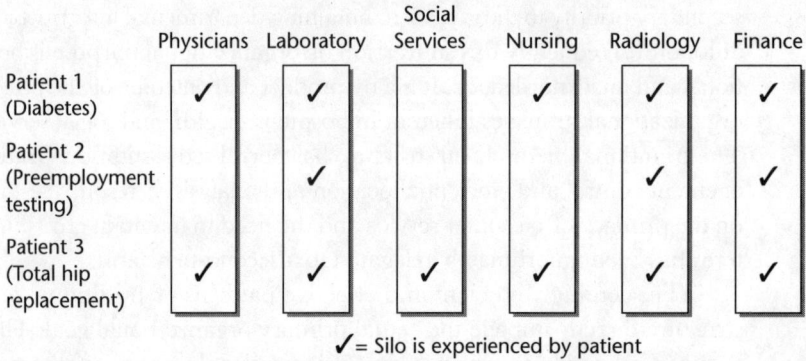

	Physicians	Laboratory	Social Services	Nursing	Radiology	Finance
Patient 1 (Diabetes)	✓	✓		✓	✓	✓
Patient 2 (Preemployment testing)		✓			✓	✓
Patient 3 (Total hip replacement)	✓	✓	✓	✓	✓	✓

✓ = Silo is experienced by patient

another department such as nursing. This flow is depicted in Figure 9.2. Yet patients, the customers of the organization's service, flow through the organization cross-functionally, crossing silo boundaries as they receive service. This flow is depicted by Figure 9.3. It is at these boundaries, where affinity and communication frequently is sparse, that service disruption occurs.

The deleterious impact of structural silos on organizational performance is magnified by integration initiatives. In the previously existing institution, where silos may have impeded smooth flow and created disconnects requiring time and resources to remedy, at least these speed bumps occurred in an environment where communication linkages existed, where managers were attuned to putting out fires, and processes for addressing system problems were very likely in place. The new entity affords double jeopardy for silo-based system disruptions. There is a risk that newly acquired services will not be integrated into existing systems; rather, they will be simply added on without attention to new ways of redesigning communication and interaction, thus creating additional silos. There is also the very real possibility that the integrated entities will devolve into separate mega-silos, dwarfing, but continuing to maintain, existing silos from the merged organizations.

Regardless of origin, process-based speed bumps pose significant threats to the successful outcome sought in integration efforts. There is a wealth of experience and knowledge available to refer to in avoiding these disruptions and derailments. Professional colleagues, consultation resources, financial and legal advisers, and scholarly publications are all resources to tap before and during the process of crafting an integrated delivery system. The old adage, "Marry in haste, repent at leisure," is well applied here. Time spent in planning a thought-

FIGURE 9.2. SILO-BASED FLOW OF INFORMATION.

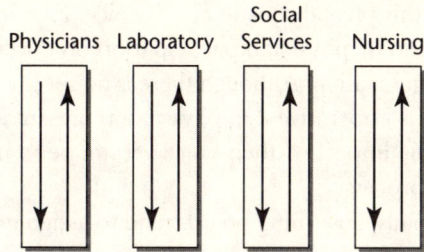

FIGURE 9.3. CROSS-FUNCTIONAL CLIENT EXPERIENCE.

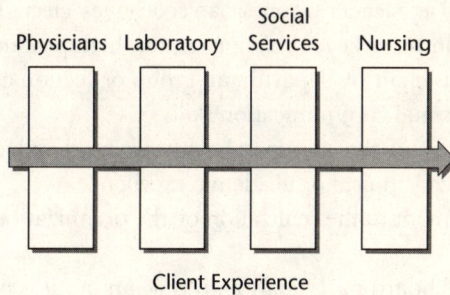

ful integration process is a good investment when weighed against the consequences of rushing in without appropriate planning preparation.

Physicians and Integration

The pivotal role of physicians in the integration of healthcare systems cannot be underestimated. Failure to attend to physician needs and input can spell disaster for the overall outcome. Although the purpose of the VIHS is to deliver more effective, efficient healthcare and social services to clients and patients, it is primarily physicians who manage access to the entry points of the medical care components of the system. Thus, physicians are both primary customers and suppliers of customers. Clearly a failure to include their input is dangerous as well as shortsighted.

Physician input and commitment can be obtained through a variety of avenues; indeed multiple routes for input are advisable to ensure optimal access and attention to this pivotal constituency's voice. First in significance and symbolism is the inclusion of physician participation on governance structures within the emergent organization. Although there is no specific ratio of physician trustees to be sought, it is imperative that physician representation be of significant volume to indicate the import of their voice yet not be so numerous as to dominate the governance process.

Additionally, governing boards tend to assign greater credibility and authority to physician input than to the voices of other constituencies represented on the board. It is important, therefore, to select physicians as trustees using criteria beyond mere membership on the medical staff. Selection criteria for a physician board member should include the following qualities:

- Perceived as a leader by physician colleagues (elected leadership position on staff)
- High-volume or key admissions status, high productivity
- Service as chair of department, group or section chief
- Interpersonal communication skills
- Knowledge of the changing healthcare marketplace
- Recognized clinical or academic excellence
- Commitment to the realization of the organization's vision

Beyond board participation, physician input is critical to the overall design and implementation planning of the integration effort. Participation by physicians on committees, task forces, and design teams is imperative. Appointed physician corporate leaders need to participate in and be capable of contributing to the evolution of the needs of the organization. Additional routes for obtaining physician input and guidance include surveys, focus groups, interviews, and questionnaires. Regardless of methodology selected, continuous physician input and reflection is crucial to successful organizational structure and implementation.

As healthcare becomes truly more health oriented and less focused on illness care, it is probable that the role of physicians as gatekeepers to system access will diminish. This decreased prominence could generate resistance and anxiety for medical staff in newly evolving systems. Yet the critical role of physicians in ensuring the success of system integration still prevails and must be respected to achieve success.

CHAPTER TEN

THE FINAL ANALYSIS

The previous chapters have addressed the scope of services, external expectations, and the internal organizational climate. Certainly, important benefits can be realized through these separate analyses. The purpose of this chapter is to describe an approach to consolidating the analyses to develop combined priority scores for the many different criteria. This final analysis will help administrative and clinical leaders prioritize efforts from the perspective of the organization as a whole.

The Balanced Scorecard

The concept of a balanced scorecard is being used in many industries now as a formal recognition that single-focus measures are inadequate. We have all known of organizations that become so focused on short-term profitability that they lose sight of their customers and their future business development. Many healthcare organizations are behaving that way in today's environment with its overriding emphasis on cost reductions. The newspapers in all major U.S. metropolitan areas regularly run stories about one medical center or another that is announcing cuts of many hundreds of staff and millions of dollars from their budgets. There is almost a competition as to which medical center can announce and implement the largest cost reduction. Yet we must retain a balanced focus of our efforts to avoid sacrificing quality and service to the members of the communities we serve.

Kaplan and Norton (1996), for example, advocate specific objectives, measures, targets, and initiatives related to four different aspects of a business: customers, internal business processes, financial, and learning and growth.

Although cost reductions in many cases are appropriate because costs have been higher than necessary, we remind you of one important fact: no organization has ever cut itself to greatness. An overemphasis on cost cutting can lead to a downward death spiral. First, large cuts are often made quickly, typically without redesigning the processes by which work is accomplished. This leads to a deterioration of service and employee morale because people are still doing their jobs in the same way, but with fewer staff and often with additional responsibilities. Poor service leads to patients' choosing to go elsewhere. This leads to more cuts, and so on.

The balanced scorecard is used to identify the key measures of critical success factors and their relative weights. There will certainly be continuing debate about the relative weights of quality, long-term outcomes, customer satisfaction, cost-effectiveness, the working environment, competitive position, the health status of the population, and other measures. However, data will be required to reach appropriate decisions. We must look at the outcomes of different practices and protocols over a longer period of time. The measures currently used by hospitals are important but very short term in nature. They seldom extend more than thirty days past hospital discharge and seldom address functionality measures related to how patients feel (e.g., pain) and how well they can function in their normal lives. When are coronary artery bypass graft surgery, balloon angioplasty, clot-dissolving agents, and continuing medical management most appropriate, and for what situations? The services, settings, and resources used will vary substantially, as will the costs and timing of those resources. The appropriate treatment is not just a matter of short-term cost; longer-term outcome and functionality data are also required.

Considerations for Combined Analyses

Before jumping into the approach and use of the tool to complete the combined analyses, we offer the following considerations and cautions. As with any other evaluation tool, the interpretation and use of the results carry certain cautions related to the underlying assumptions and analysis methodology.

Level of Detail

As with any other structured analysis tool, these analyses of a healthcare organization or VIHS can be undertaken at many different levels of detail. We suggest beginning with an overview analysis first, to raise and prioritize the major issues to be addressed. Doing extremely detailed analyses may be of little value and financially wasteful if there are major voids or deal killers.

Types of Analyses Not Included in the Model

We specifically acknowledge that the following analyses are not emphasized in this evaluation model:

- *Financial audit.* A separate financial audit is necessary when crafting an integrated health system. We chose to exclude it from this book for two reasons. First, a health system should be built around the health and healthcare needs of the population, while ensuring financial viability and profitability. Second, most mergers and network formations have historically overemphasized the financial aspects.
- *Legal analyses.* Separate legal analyses are clearly necessary when crafting new or revised virtual health systems. This is an area where using consultants with specific expertise in forming large affiliations, mergers, networks, and systems is important.

The VIHS should be driven by the requirements of the population served, using the highest-quality, most cost-effective configuration of foci, settings, and resources available. The financial and legal analyses should support the efforts to the extent possible, within the financial and legal constraints.

Other types of analyses may be required while integrating selected aspects of a VIHS. Environmental impact statements, for example, may be required for certain facility changes, and agreements will be required among collaborating but separately owned VIHS components.

Number of Criteria

The number of dimensions and number of criteria within each dimension will affect the final scores. If the final weighted scores are added, then the final scores, and relative emphasis, of scope of services, external expectations, and internal organizational climate will be affected by the number of criteria within each. Alternatively, you can divide by the number of criteria within each section or dimension to arrive at an average score for each. Both approaches have advantages and disadvantages. You simply have to understand which you are using and why.

Causes of Most Failures

Although initial analyses often stress the financial, legal, and operational implications of merging or affiliating organizations, failures are most commonly caused by culture and human dimension issues within the internal organizational climates of the organizations involved. Hence, we want to emphasize that any evaluation must include formal consideration of these issues.

Tailoring the Assessment Tools

Organizations and situations vary so widely that it is impossible to develop tools that meet the specific needs of all organizational situations. Thus you should see this as a template, or guide, to critical issues. It is important to add criteria specific and significant to your situation.

Assigning Weights

Assign weights, to the extent practical, based on the needs of the population, and not on your current organizational situation. Our advice is to start with the population-based needs and then work backward to the alternatives to address those needs, and then to your current situation. Your weights should also reflect the expectations of accreditation and licensing organizations. You should not eliminate or minimize critical external review criteria.

Underemphasis on Deficiencies

Be cautious of placing very small weights on the foci, settings, and resources for which you have current deficiencies. Otherwise you may underemphasize those areas and consequently miss opportunities. Carried to an extreme, if you were to weight everything you are currently doing high, and everything you are currently not doing as zero, then you would end up with what you have now.

Scoring

An example of a combined analysis is illustrated in Figure 10.1. Less detailed criteria are used in this figure than those discussed in Chapters Four, Six, and Eight for the purpose of describing how weights are used to arrive at a consolidated score. The scoring can be handled in different ways. Following are some useful approaches:

- A tiered or roll-up process in which the detailed data from the analyses in Chapters Four, Six, and Eight are summarized, and then only the broader categories are included in the combined analyses later in Figure 10.1.
- Blended scores of the different criteria and scores.
- Low scores, to focus attention on the areas that score lowest. However, this approach, as used by the JCAHO, misses the overall performance.

- Differences of scores among units. This approach focuses most attention on the criteria with the greatest variation and may be particularly useful when assessing the culture, human dimensions, and organizational climates among VIHS components.

Alternatively, you can create one very large spreadsheet that includes all the detail. This is clearly the most complete. However, if everything is included in one large spreadsheet, it is easy for leaders to get bogged down in details and miss the important relative priorities.

Simulations

Simulate your assessment with different weights. This can be done very easily with a computer spreadsheet and will provide insight of how sensitive your decisions are to your choices of weights. As weights of the different dimensions and criteria are changed in the spreadsheet, you can observe changes in total scores and, more important, the relative priorities of different potential actions.

Reducing Bias

Ensure that people with a wide range of knowledge, perceptions, and interests participate in the assessment. This avoids missed priorities and opportunities due to lack of knowledge or limited scope of perceptions. Using multiple sources of information is one approach to reduce biases and consequences of inadequate knowledge about the whole situation.

Combined Analyses for a VIHS or Organization

In this section we illustrate how the measures from the previous chapters can be combined into a single evaluation tool. The example evaluation tool, illustrated in Figure 10.1, will be used to describe the use of the relative scoring system to develop a balanced scorecard.

Type of Analysis Weights

The first step is to determine the relative emphasis to be placed on each type of analysis or section of the combined analysis. All are important but emphasize different aspects of your organization or VIHS. Important cautions are described later in this chapter. The type of analysis weights are entered in column a of

FIGURE 10.1. A COMBINED ANALYSIS FOR THE ORGANIZATION OF A VIRTUAL INTEGRATED HEALTH SYSTEM.

Type of Analysis		Important Criteria	Combined Wt.	Criteria Scores (0–4, 4=high)			Weighted Scores			Comments
Number	Name / Sect. Wt. (0–4) a	Description / Crit. Wt. (0–4) b	$c=(a*b)$	Your Org. d	Org. A e	Org. B f	Your Org. $g=(c*d)$	Org. A $h=(c*e)$	Org. B $i=(c*f)$	
1	Scope of Services [4]									
		Social and Environmental Conditions [4]	16	1			16	0	0	Risky behaviors unaddressed
		Health or Clinical Conditions [4]	16	4			64	0	0	
		Foci [4]	16	3			48	0	0	
		Settings [4]	16	2			32	0	0	Community settings unaddressed
		Core/Key Processes (included in external expectations below)								
		Resources [4]	16	3			48	0		
		Other (other priority criteria identified by organization) [ ]	0				0	0		
2	External Expectations [3]									
		Leadership and Governance [4]	12	2			24	0	0	
		Strategic Planning [4]	12	3			36	0	0	
		Human Resources Management and Development [4]	12	3			36	0		

Type of Analysis and Criteria Weights (to fill in boxed areas in columns a and b)

0 = No current weight or value, or not applicable
1 = Minimal weight or value
2 = Modest weight or value
3 = Significant weight or value
4 = Important weight or value

Criteria Scores (to fill in the boxed areas in columns d, e, and f):

0 = No competence, coverage, or success, or not applicable
1 = Minimal competence, coverage, or success
2 = Partial competence, coverage, or success
3 = Significant competence, coverage, or success
4 = Substantial competence, coverage, or success

FIGURE 10.1.   A COMBINED ANALYSIS FOR THE ORGANIZATION OF A VIRTUAL INTEGRATED HEALTH SYSTEM. (continued)

Type of Analysis		Important Criteria	Com-bined Wt.	Criteria Scores Score (0–4, 4=high)			Weighted Scores			Comments
Number	Name	Description		Your Org.	Org. A	Org. B	Your Org.	Org. A	Org. B	
	Sect. Wt. (0–4) a		Crit. Wt. (0–4) b	d	e	f				
			c=(a*b)				g=(c*d)	h=(c*e)	i=(c*f)	
		Process and Quality Improvement	12	2			24	0	0	Fragmentation
		Information Planning and Management	12	2			24	0	0	Implementation issues
		Continuum of Care	12	1			12	0	0	Access issues
		Client/Patient Rights and Satisfaction	12	3			36	0	0	
		Prevention and Education	12	3			36	0	0	
		Managing the Environment	12	4			48	0	0	
		Other (other priority criteria identified by organization)	0				0	0	0	
3	**Internal Organizational Climate** [4]	Structural Fluidity	16	1			16	0	0	Poor bias for action
		Measurement	16	2			32	0	0	Dearth of outcome measures
		Leadership	16	2			32	0	0	Quality of work life
		Paradoxical	16	3			48	0	0	
		Deep Culture	16	3			48	0	0	
		Stretch Goals	16	3			48	0	0	
		Execution with Excellence	16	3			48	0	0	
		Other (other priority criteria identified by organization)	0				0	0	0	

Combined Weighted Score (sum of weighted scores)   765   0   0

Figure 10.1 for scope of services, external expectations, and internal organizational climate.

An infinite number of possible scoring methods and scales can be used. However, we recommend a simple scoring system initially. The following scores are used for the type of analysis weights and the criteria weights, used in the following discussions:

0   No current weight or value, or not applicable

1   Minimal weight or value

2   Modest weight or value

3   Significant weight or value

4   Important weight or value

Your organization may choose a different scale, but be cautious not to make it overly complex. The purpose is to prioritize, not develop some exact score.

These scores are later used to develop a combined weight for the criteria.

Criteria Weights

Sample criteria for each dimension of each of the three types of analyses are listed in Figure 10.1. The criteria are abbreviated from the more complete lists described in Chapters Four, Six, and Eight. Depending on the level of detail desired, the combined analysis table may include summary categories or a complete listing of all criteria.

The relative weights for the criteria are entered in the boxes in column b of Figure 10.1. The same simple 0–4 scale is suggested for the criteria weights as used for the type of analysis weights. Space is allowed in Figure 10.1 to enter additional criteria identified by your organization. You may want to create one or more spreadsheets that uniquely meet the needs of your organization.

Readers familiar with the JCAHO will recognize that its scoring system is the reverse of ours. We chose to use larger numbers to indicate better scores, because this is the most typical and intuitive approach.

Combined Weights

The combined weights are calculated in column c of Figure 10.1, as the multiplication of the weights for the type of analysis and criteria (a × b). The combined weights are intermediate calculations that are later multiplied by the criteria scores to arrive at the final weighted scores.

Criteria Scores

Columns d, e, and f of Figure 10.1 allow you to enter the scores for how well each of the criteria is addressed in your organization and in two other organizations. The other organizations may be VIHS partners or competitors, depending on the comparisons you want to make. The following scores are used for the criteria, as described in Chapters Four, Six, and Eight:

0   No competence, coverage, or success, or not applicable

1   Minimal competence, coverage, or success

2   Partial competence, coverage, or success

3   Significant competence, coverage, or success

4   Substantial competence, coverage, or success

Weighted Scores

Weighted scores are calculated for your organization and any comparison organizations for which you have estimated criteria scores. The weighted scores are the multiplication of the respective combined weight from column c and the criteria score from column d, e, or f, respectively. The totals are entered in columns g, h, and i, respectively. These weighted scores then give the relative emphasis placed on each of those criteria, given your judgments of the relative weights of the type of analysis and criteria. One of the real strengths of the consolidated analysis is the dialogue regarding the relative importance of different aspects of the organization's business.

A Combined Analysis

To illustrate the calculations, an example for the combined analysis is illustrated in Figure 10.1. For illustration, your organization is compared to two other organizations: Organization A and Organization B.

The scope of services and internal organizational climate were both given a weight of 4, indicating important weight or value. External expectations were given a weight of 3, indicating that this category is weighted lower than the other two for the current analyses.

Weights from 0 to 4 were assigned to each of the criteria. There are several points to note in this example. First, a single set of weights is assigned to the criteria for all organizations compared. The scores on those criteria, of course, are likely to be different for each organization. Second, this example assumes the tiered approach to the criteria related to the scope of services. The scoring of the criteria in Figure 10.1 only makes judgments about the extent to which those criteria are addressed. The more detailed analyses of the criteria within each of the six dimensions (social and environmental conditions, health or clinical conditions, foci, settings, core/key processes, and resources) would be scored separately, as described in Chapter Four.

Each of the criteria is then scored for each of the three organizations included in the comparison. Depending on the level of detail of these assessments, the scores may be simple judgments or may be based on data gathered about each organization. For the results of the analysis to be meaningful, the scoring should be as objective as possible. It is easy to score your own organization high on everything subjectively, but by doing this, you gain little from the assessment process. In this example, none of the three organizations addressed crime, violence, or income in the communities they served, so they received scores of zero.

The scores of these analyses are relative and have no value in absolute terms. The combined analysis will identify areas for further work and emphasis.

CHAPTER 11

PLANNING VIRTUALLY INTEGRATED HEALTH SYSTEMS

Now that the analyses process is complete, what do you do with the information gleaned? Figure 10.1 provided an illustration of identified areas of opportunity or limitation that may need to be addressed. The task becomes determining what criteria, if any, are high priority and merit significant, prompt attention, what can be delayed, what can be delegated, and what can be ignored.

It is important to determine first if the deficiency merits immediate attention. If the answer is a definitive yes, a more detailed evaluation is justified prior to action. For example, if the deficient score is focused on an organizational inability to take appropriate and prompt action, a more thorough evaluation of the locus of the deficiency is needed prior to the design or implementation of remedies. Again, if you do not know your destination, any road will take you there.

Determination of the gravity of the deficiency and its impact on the organization can be assessed in several ways. The following questions to ask are key:

- What is the impact of this issue on overall organizational performance?
- How does this deficiency affect the organization's ability to consider affiliation, mergers, and alliances (or decrease the attractiveness of a potential alliance if the assessment is being completed in anticipation or evaluation of potential partners)?
- What is the opportunity cost of the deficiency?
- How does the deficiency, duplication, gap, or overlap affect the customer?

Using the bias for action as an assessed area of concern (see Figure 11.1), these questions would result in discussion focused on the following sample responses to the questions:

- Have new and needed programs or initiatives been delayed or deferred because of slow action or no action?
- When external competitive threats have been realized, was the organization able to respond in an effective and timely manner?
- What is the analysis cycle time for key decisions?
- How elaborate and convoluted is the approval process for routine decisions (hiring staff, capital acquisitions below an approval ceiling, and so on)?
- How many management levels are there between the CEO and the front line of service to the customer?
- Are external consultants overused? Are they hired for projects that staff could complete effectively?
- Is the financial control and audit process overly cumbersome and stifling?

The summary response to these questions determines where and how the organization addresses key integration initiatives. Planning to do so requires further design attention. The planning format in Figure 11.2 may serve as a guide in addressing these issues. This format is useful in a variety of situations. Continuing the example of addressing assessed deficiencies in the dimension of structural fluidity and noting concerns about bias for action, the plan shown in Figure 11.3 could be posed.

The emergence of a national system with a focus on the health of a population instead of individual illness and with universal access may never come to pass in the United States, at least not with the existing pervasive culture of individualism. The United States is the only country of which we are aware that is attempting to form a national health insurance culture after the advancements in medicine and technology have surpassed anything we would previously thought possible. Increasing technology and specialization have created many of the problems we now see with any attempt at system reform. Proposals and changes cannot progress very far without running into a discussion about ethics and values and individual versus societal and community needs. The problem we face with changing the social services and healthcare in America is that change requires asking most people to give up something.

Health and social services for everyone is a value judgment. As a society, we must respond to two important questions: Is there really a problem with the American healthcare system and does it justify a public solution? How will we respond along ideological and political party lines? To some people, 37 million uninsured

FIGURE 11.1. ASSESSED DEFICIENCY IN STRUCTURAL FLUIDITY.

Type of Analysis		Important Criteria		Com-bined	Criteria Scores (0–4, 4=high)				Weighted Scores			Comments
Number	Name	Sect. Wt. (0–4) a	Description	Crit. Wt. (0–4) b	Wt. c=(a*b)	Your Org. d	Org. A e	Org. B f	Your Org. g=(c*d)	Org. A h=(c*e)	Org. B j=(c*f)	

Type of Analysis and Criteria Weights (to fill in boxed areas in columns a and b):
- 0 = No current weight or value, or not applicable
- 1 = Minimal weight or value
- 2 = Modest weight or value
- 3 = Significant weight or value
- 4 = Important weight or value

Criteria Scores (to fill in the boxed areas in columns d, e, and f):
- 0 = No competence, coverage, or success, or not applicable
- 1 = Minimal competence, coverage, or success
- 2 = Partial competence, coverage, or success
- 3 = Significant competence, coverage, or success
- 4 = Substantial competence, coverage, or success

Number	Name	a	Description	b	c	d	e	f	g	h	j	Comments
3	Internal Organizational Climate	[4]	**Structural Fluidity** Measurement	4	16	1			**16**	0	0	**Poor bias for action**
				4	16	2			32	0	0	Dearth of outcome measures
			Leadership	4	16	2			16	0	0	Quality of work life
			Paradoxical	4	16	3			48	0	0	
			Deep Culture	4	16	3			48	0	0	
			Stretch Goals	4	16	3			48	0	0	
			Execution with Excellence	4	16	3			48	0	0	
			Other (other priority criteria identified by organization)	0					0	0	0	

FIGURE 11.2. PLANNING OUTLINE.

Planning Outline

Mission: _____

Values: _____

Vision: _____

Dimension Goal	Criterion Objective	Activity Benchmark	Responsible Party	Timetable	Review Status

FIGURE 11.3. PLANNING OUTLINE WITH BIAS FOR ACTION EXAMPLES.

Dimension Goal	Criterion Objective	Activity Benchmark	Responsible Party	Timetable	Review Status
Structural Fluidity	Bias for Action	Three levels of management between CEO and frontline	Senior leadership team	3 months	
		Reduce cycle time for decision analysis to 5 days	Chief financial officer	1 month	
		Vacant positions filled without approval	Vice president of human resources	1 month	
		Unit/department managers have capital expenditure approval to $5,000	Senior leadership team	2 months	
		Inventory of consultants/projects	Vice president of operations	2 weeks	

in a nation with the finest healthcare in the world is appalling. To others, it may not look so bad, considering the fact that 37 million is only 14 percent of the U.S. population. Is this really a problem that requires a public policy response? Solutions—particularly policy solutions—depend on voters. Voters respond to the framing of the problem, the presence of exciting leaders and interest groups, and their own feelings concerning the acuity of the problem and how the solution will affect them, especially financially. Considerations such as these cause major credibility problems for any significant healthcare reform initiatives.

We must decide how we will allocate and control costs. Should we continue to allow people to go without necessary social services or medical coverage, or have lower standards for their health? The results of the polls and the underlying indecision appear to reflect a pervasive distrust of the government's ability to administer social programs effectively.

Universal access and affordability cannot be attained unless the growth in services and medical costs is slowed significantly and brought under control. It started with the emergence of a prospective payment system and continues with the integration of healthcare and social services organizations. Society is spending too much on healthcare and for the wrong things. There appears to be a disproportion in relation to other societal needs (Lamm, 1990).

No country provides unlimited healthcare and social services access to all of its citizens. The value judgments and rationing process that invariably come with a national health product were made before the huge advancements in medicine and technology and before the swelling of elderly and disadvantaged populations. The decisions of other countries to ration care and services occurred before there was the ability to transplant organs, or separate Siamese twins, or prolong life longer than is meaningful. In the United States, this reform means taking away the technological access in healthcare we have become used to. It means reallocating the cost for illness care toward improving the societal conditions that provide the patients for acute and chronic care organizations—true demand management. To our knowledge, no other country in the world is both willing and able to provide access to all technologies for all society.

We hope the insights, approaches, and tools we have provided in this book will help organizations avoid many of the problems, or speed bumps, on the road to integration. Through the three aspects of analyses—scope of services, external expectations, and culture and human dimensions—either separate or combined, we hope you will identify issues early by using this tool. Patients, clients, and consumers are knowledgeable and critical. They are cognizant of both quality and cost, and they have high expectations of the organizations that deliver their healthcare.

Just when we begin to think we have done what we need to do in terms of our healthcare, health, and social systems, another issue will emerge. The focus will shift with new technology, and the market we serve within these systems will no longer be a county, a state, or even a single country; it will be the world. Very soon the issues of access, integration, costs, reimbursement, and more will involve the development of a true virtually integrated health system.

◆ ◆ ◆

This book is a work in progress. We have attempted to provide useful insights, approaches, and tools to help your organization or system analyze its situation and make good decisions. However, we must continue to improve. Therefore, we invite ideas, questions, examples, and ideas to improve and expand the approaches and tools in this book.

APPENDIX A

CROSSWALK AMONG
EXTERNAL REVIEW BODIES

This appendix contains a set of crosswalks between several external review bodies' standards and the Malcolm Baldrige National Quality Award criteria. The organizations crosswalked are the Joint Commission on Accreditation of Healthcare Organizations (JCAHO), the National Council on Quality Assurance (NCQA), the Council on Accreditation (COA), and the Commission on Accreditation of Rehabilitation Facilities (CARF). All are crosswalked initially against the Baldrige criteria. These crosswalks are high level, meaning they crosswalk only dimension against dimension rather than standard by standard. The standard-by-standard crosswalks are too lengthy for inclusion here.

Your organization or virtually integrated health system may want to rename, add, or delete one or more of the dimensions. In addition, it is possible to expand one of the dimensions into two or more dimensions. We advise caution in eliminating or changing the nine key dimensions of external review bodies because this strategy may cause you to omit potentially important dimensions and criteria from your evaluation.

FIGURE A.1.   CROSSWALK BETWEEN BALDRIGE HEALTHCARE DIMENSIONS AND COA DIMENSIONS.

Baldrige Dimensions	Agency in the Community	Agency Governance and Administration	Personnel	Quality Assurance	Fiscal Management	Facilities and Equipment	Intake and Assessment	Service Planning	Implementation of Service Planning Termination and Aftercare	Client Information and Technology
Leadership	X	X		X	X	X				X
Information and Analysis				X	X		X	X	X	
Strategic Planning	X	X			X					
Human Resource Development and Management			X							
Process Management				X			X	X	X	
Performance Results	X			X				X	X	
Focus on Patient/ Stakeholder Satisfaction	X			X			X	X	X	X

FIGURE A.2. CROSSWALK BETWEEN BALDRIGE HEALTHCARE DIMENSIONS AND NCQA DIMENSIONS.

Baldrige Dimensions	NCQA Dimensions					
	Quality Management and Improvement	Utilization Management	Credentialing	Member Rights and Responsibilities	Preventative Health Services	Medical Records
Leadership	X	X	X	X	X	
Information and Analysis	X	X		X	X	X
Strategic Planning	X	X				
Human Resource Development and Management	X	X	X			
Process Management	X	X	X	X	X	
Performance Results	X	X	X	X	X	X
Focus on Patient/ Stakeholder Satisfaction	X	X		X	X	X

FIGURE A.3.   CROSSWALK BETWEEN BALDRIGE HEALTHCARE DIMENSIONS AND JCAHO DIMENSIONS.

Baldrige Dimensions	JCAHO Network Dimensions							
	Rights, Responsibilities, and Ethics	Education and Communication	Continuum of Care	Health Promotion and Disease Prevention	Leadership	Management of Human Resources	Management of Information	Improving Network Performance
Leadership	X				X			X
Information and Analysis					X		X	X
Strategic Planning					X		X	X
Human Resource Development and Management			X		X		X	X
Process Management		X	X	X	X		X	X
Performance Results			X	X	X	X	X	X
Focus on Patient/ Stakeholder Satisfaction	X	X			X		X	X

FIGURE A.4. CROSSWALK BETWEEN BALDRIGE HEALTHCARE DIMENSIONS AND CARF DIMENSIONS.

Baldrige Dimensions	CARF Dimensions				
	Organizational Quality	Organizational Performance	Accessibility, Safety, and Transportation	Quality Services and Treatment	Program Standards
Leadership	X	X	X		X
Information and Analysis	X	X		X	X
Strategic Planning					X
Human Resource Development and Management				X	X
Process Management		X	X	X	X
Performance Results		X		X	X
Focus on Patient/ Stakeholder Satisfaction	X		X	X	X

FIGURE A.5. CROSSWALK BETWEEN JCAHO DIMENSIONS AND NCQA DIMENSIONS.

JCAHO Dimensions	NCQA Dimensions					
	Quality Management and Improvement	Utilization Management	Credentialing	Member Rights and Responsibilities	Preventative Health Services	Medical Records
Rights, Responsibilities, and Ethics		X		X		
Education and Communication		X		X	X	
Continuum of Care	X	X	X	X		
Health Promotion and Disease Prevention	X				X	
Leadership	X	X	X	X	X	
Management of Human Resources	X	X	X			
Management of Information	X			X		X
Improving Network Performance	X	X	X	X	X	

FIGURE A.6. CROSSWALK BETWEEN NCQA DIMENSIONS AND COA DIMENSIONS.

NCQA Dimensions	COA Dimensions									
	Agency in the Community	Agency Governance and Administration	Personnel	Quality Assurance	Fiscal Management	Facilities and Equipment	Intake and Assessment	Service Planning	Implementation of Service Planning, Termination, and Aftercare	Client Information and Technology
Quality Management and Improvement	X	X		X	X	X				
Utilization Management				X	X		X	X	X	
Credentialing	X		X	X						
Member Rights and Responsibilities	X	X		X						
Preventative Health Services	X	X		X						
Medical Records										X

FIGURE A.7.   CROSSWALK BETWEEN JCAHO DIMENSIONS AND CARF DIMENSIONS.

JCAHO Dimensions	CARF Dimensions				
	Organization Quality	Organization Performance	Access, Safety, and Transportation	Quality Services and Treatments	Program Standards
Rights, Responsibilities, and Ethics	X		X	X	X
Education and Communication				X	X
Continuum of Care		X		X	X
Health Promotion and Disease Prevention				X	
Leadership	X		X		
Management of Human Resources	X				
Management of Information	X	X		X	X
Improving Network Performance		X	X	X	X

FIGURE A.8. CROSSWALK BETWEEN JCAHO DIMENSIONS AND COA DIMENSIONS.

JCAHO Dimensions	COA Dimensions									
	Agency in the Community	Agency Governance and Administration	Personnel	Quality Assurance	Fiscal Management	Facilities and Equipment	Intake and Assessment	Service Planning	Implementation of Service Planning, Termination, and Aftercare	Client Information and Technology
Rights, Responsibilities, and Ethics	X	X		X		X	X			X
Education and Communication	X									
Continuum of Care	X						X	X	X	
Health Promotion and Disease Prevention	X	X								
Leadership		X			X	X				
Management of Human Resources			X						X	
Management of Information				X				X		
Improving Network Performance				X						

APPENDIX B

COMPREHENSIVE ASSESSMENT TOOL

This appendix contains a comprehensive assessment tool from Chapters Four, Six, and Eight. This has been included for the convenience of readers, so the tables can be extracted and used for discussions of the appropriate dimensions and criteria for each of the three major types of analyses: scope of services (Figure B.1), external expectations (Figure B.2), and internal organizational climate (Figure B.3). Your organization or virtually integrated health system may want to add, amend, or delete particular criteria. We advise caution in totally eliminating key dimensions, because you may omit from your analyses potentially important opportunities for your organization or VIHS.

Spreadsheets can be created for these analyses or are available from the authors. One good reason to put the analyses in computer spreadsheets is to allow your organization to simulate the impacts of different weights and criteria scores. The spreadsheets can be used to simulate the combined weights and priorities based on perceptions of different groups. The relative priorities can then be compared to see if there is any qualitative difference in the things prioritized and actions taken. Remember that the scores are relative; they have no specific meaning in absolute terms. All of the scores could be doubled, for example, if larger scores were allowed, but the relative priorities will be the same.

FIGURE B.1. ANALYSIS OF SCOPE OF SERVICES.

Major Section		Important Criteria	Criteria Scores Score (0–4, 4 = high)				
Number	Name	Description			Your Organization		Comments
		a	b	c	d	e	f

Scoring of criteria (fill in the boxed areas):

0 = No competence, coverage, or success, or not applicable
1 = Minimal competence, coverage, or success
2 = Partial competence, coverage, or success
3 = Significant competence, coverage, or success
4 = Substantial competence, coverage, or success

1 **Scope of Services**

Social and Environmental Conditions
Environmental pollution
Crime and violence
Community and social support
Family and living situation
Education and vocational levels
Employment and income levels
Risk factors and behaviors
Other priority conditions, identified by organization

Health or Clinical Conditions
Diseases and injuries (seventeen categories)
Operations (sixteen categories)
Health status and contact with health services (eight categories)
Causes of injury and poisoning (twenty-two categories)
Maintenance and enhancement of health

FIGURE B.1. ANALYSIS OF SCOPE OF SERVICES. (continued)

Major Section		Important Criteria	Criteria Scores Score (0–4, 4 = high)			
Number	Name	Description		Your Organization		Comments
	a		b	c	d	e	f	

Other priority conditions, identified by organization

Foci
Promotion
Protection
Prevention
Detection
Diagnosis and assessment
Treatment
Habilitation and rehabilitation
Maintenance
Hospice
Support
Advocacy
Education
Research
Enabling
Other priority foci, identified by organization

Settings
Areawide
Community
Houses of worship

FIGURE B.1. ANALYSIS OF SCOPE OF SERVICES. (continued)

Major Section		Important Criteria		Criteria Scores Score (0–4, 4 = high)					Comments
Number	Name	Description			Your Organization				
	a		b	c	d	e	f		
		Schools							
		Work							
		Mobile							
		Home							
		Ambulatory							
		Partial-day care							
		Inpatient							
		Freestanding support							
		Other priority settings, identified by organization							

Core/Key Processes (may score in external expectations)

Leadership and governance					
Strategic planning					
Human resources management and development					
Process and quality improvement					
Information planning and management					
Continuum of care					
Client/patient rights and satisfaction					
Prevention and education					
Managing the environment					

FIGURE B.1. ANALYSIS OF SCOPE OF SERVICES. (continued)

Major Section		Important Criteria			Criteria Scores Score (0–4, 4 = high)			Comments
Number	Name	Description				Your Organization		
	a		b	c	d	e	f	

Other priority criteria, identified by organization

Resources
  Human resources
    Physicians—specialists
    Physicians—generalists
    Physicians—residents
    Physician assistant/
      nurse practitioner
    Nurses
    Pharmacists
    Therapists
    Technicians and allied health
    Medical assistants
    Office staff
    Professional /administrative
    Trades
    Service/maintenance
    Community health nurses/workers
    Child/family welfare workers
    Teachers and coaches
    Clergy
    Influential community members
    Patient/client and family

FIGURE B.1. ANALYSIS OF SCOPE OF SERVICES. (continued)

Major Section		Important Criteria			Criteria Scores Score (0–4, 4 = high) Your Organization			Comments
Number	Name	Description	b	c	d	e	f	
	a							
		Other human resources						
		Physical resources						
		Facilities						
		Equipment (particularly expensive/specialized equipment)						
		Instruments						
		Supplies						
		Other physical resources						
		Financial resources						
		Organizations						
		Information						
		Other priority criteria, identified by organization						

FIGURE B.2. ANALYSIS OF EXTERNAL EXPECTATIONS.

Major Section		Important Criteria	Criteria Scores Score (0–4, 4 = high)			
Number	Name	Description	Your Organization			Comments
	a	b	c	d	e	f

Scoring of criteria (fill in the boxed areas):
0 = No competence, coverage, or success, or not applicable
1 = Minimal competence, coverage, or success
2 = Partial competence, coverage, or success
3 = Significant competence, coverage, or success
4 = Substantial competence, coverage, or success

2 External Expectations

Leadership and Governance
Communications
Community involvement
Establishment and alignment of:
  Mission
  Vision
  Values
  Critical success factors
  Goals
  Key processes
  Key indicators
  Feedback
Other priority criteria, identified
  by organization

Strategic Planning
Current financial status
Financial objectives
Customer requirements
Customer expectations
Resource allocation
Staff requirements

FIGURE B.2. ANALYSIS OF EXTERNAL EXPECTATIONS. (continued)

Major Section		Important Criteria			Criteria Scores Score (0–4, 4 = high)			Comments
Number	Name	Description			Your Organization			
	a		b	c	d	e	f	
		Staff expectations						
		Environmental changes						
		Organizational capabilities						
		Future risks						
		New services						
		Services redesign						
		Other priority criteria, identified by organization						
		Human Resources Management and Development						
		Alignment with organizational goals						
		Staffing (right number of people)						
		Job categories (right kind of people)						
		Skills assessments						
		Competency evaluations						
		Credentialing clinicians						
		Employee satisfaction						
		Employee development						
		Continuing education						
		Training						
		Other priority criteria, identified by organization						

FIGURE B.2. ANALYSIS OF EXTERNAL EXPECTATIONS. (continued)

Major Section		Important Criteria			Criteria Scores Score (0–4, 4 = high) Your Organization			Comments
Number	Name	Description						
	a		b	c	d	e	f	

Process and Quality Improvement
Input from:
  Clients/patients
  Staff
  Clinicians
  Vendors/contractors
Geographic settings identified
Assess external environment
Identify all important processes
Develop measures for processes
Monitor processes
Evaluate data
Implement/improve processes
Other priority criteria, identified
  by organization

Information Planning and Management
Needs assessment
Identify:
  External data requirements
  Internal data needs
Solicit input from:
  Clinicians
  Staff
  Suppliers
  Providers

FIGURE B.2. ANALYSIS OF EXTERNAL EXPECTATIONS. (continued)

Major Section		Important Criteria		Criteria Scores Score (0–4, 4 = high) Your Organization			Comments
Number	Name	Description					
	a		b	c	d	e	f

Description column contents:

Information needs for:
- Any client/patient
- Any time they enter system
- Any place they enter system
- Plan from needs assessment
- Address:
  - Confidentiality
  - Security
  - Accessibility
- Access to literature
- Coordinate medical record:
  - Numbering system
  - Patient-level data requirements
  - System access
  - Uniformity
  - Standardize operational definitions
- Other priority criteria, identified by organization

Continuum of Care
- Patient/client care
- Integrate:
  - Admission/entry
  - Registration
  - Scheduling coordination
- Assessment:
  - Care needs are assessed at all sites

FIGURE B.2   ANALYSIS OF EXTERNAL EXPECTATIONS. (continued)

Major Section		Important Criteria				Criteria Scores Score (0–4, 4 = high) Your Organization			Comments
Number	Name	Description	a	b	c	d	e	f	
		Assessments are appropriate to site							
		Level of care provided is appropriate							
		Services are appropriate							
		Services are timely							
		Integrated treatment/services:							
		Communication between sites							
		Communication between practitioners							
		Transfer between sites is seamless							
		Discharge							
		Follow-up/aftercare							
		Population needs assessment:							
		Sufficient disciplines and specialists available							
		Appropriate services available							
		Appropriate levels of care available							
		New service design							
		Existing service evaluation							
		Measure processes							
		Improve service/processes							
		Incorporate support services							
		Identify community resources							
		Incorporate community services							
		Incorporate administrative support services							
		Population eligibility for services							
		Mechanism to communicate services							

FIGURE B.2.   ANALYSIS OF EXTERNAL EXPECTATIONS. (continued)

Major Section		Important Criteria			Criteria Scores Score (0–4, 4 = high) Your Organization			Comments
Number	Name	Description						
	a		b	c	d	e	f	

Systemwide access issues:
Transportation
Hours of service
Cultural barriers
Language barriers
Other priority criteria, identified
  by organization

Client/Patient Rights and Satisfaction
Informed consent
Client/patient participate
  in decisions
Recognition of:
  Cultural differences and values
  Religious practices and values
Patient/client input gathered
Patient/client satisfaction measured
Information used for process/
  service improvement
Ethics/behavior code addressing:
  Conflicts
  Incentives
  Decision making
Mechanism for resolving problems

FIGURE B.2. ANALYSIS OF EXTERNAL EXPECTATIONS. (continued)

Major Section		Important Criteria	Criteria Scores Score (0–4, 4 = high)			Comments
Number	Name	Description	Your Organization			
	a		b	c		
			d	e	f	

Other priority criteria, identified by organization

Prevention and Education
Education needs determined for:
 Patient/client
 Family/support
 Community
 At-risk populations
Population health promotion is:
 Assessed
 Planned
 Implemented
Disease prevention is:
 Identified
 Assessed
 Planned
Community resources are included
Barriers are identified:
 Economic
 Family/social
 Language
 Access
Aftercare is:
 Assessed
 Tracked
 Monitored

FIGURE B.2. ANALYSIS OF EXTERNAL EXPECTATIONS. (continued)

Major Section		Important Criteria	Criteria Scores Score (0–4, 4 = high)			Comments
Number	Name	Description		Your Organization		
	a	b	c	d	e	f

Other priority criteria, identified by organization

Managing the Environment
Plan development incorporates:
  Building
  People
  Equipment
Process for monitoring:
  Safety and infection control
  Controlling environmental hazards
  Address risks
  Accident prevention
  Building codes
  Fire codes
  Disaster plans
  Local requirements
  State requirements
  Federal requirements
  Implement improvements
Other priority criteria, identified by organization

FIGURE B.3. ANALYSIS OF THE INTERNAL ORGANIZATIONAL CLIMATE.

Major Section		Important Criteria	Criteria Scores Score (0–4, 4 = high)			Comments
Number	Name	Description	Your Organization			
	a		b	c		
			d	e	f	

Scoring of criteria (fill in the boxed areas):

0 = No competence, coverage, or success, or not applicable   3 = Significant competence, coverage, or success
1 = Minimal competence, coverage, or success   4 = Substantial competence, coverage, or success
2 = Partial competence, coverage, or success

3	Internal Organizational Climate	**Structural Fluidity**				
		Bias for action:				
		How quickly are new programs or ideas evaluated?				
		How well does the organization respond to:				
		Internal threats?				
		External threats?				
		Short analysis cycle time for decision making?				
		Easy-to-use approval process?				
		Few layers of management between front line and CEO?				
		Little use of external consultants to validate internal decisions?				
		Minimal audit control systems in place?				
		Other queries, identified by organization				
		Turnover in pointman:				
		Reasonable tenure for CEO position?				
		Multiple project leaders?				

FIGURE B.3. ANALYSIS OF THE INTERNAL ORGANIZATIONAL CLIMATE. (continued)

Major Section		Important Criteria	Criteria Scores Score (0–4, 4 = high)					
Number	Name	Description	Your Organization					Comments
	a		b	c	d	e	f	
		Deep involvement in new ventures/programs?						
		Response to new initiatives is positive?						
		Quick response time to new initiatives?						
		Organization embraces new technology?						
		Active recruitment of leadership?						
		Extensive leadership development?						
		Other queries, identified by organization						
		Teams/taskforces:						
		Many new project launches?						
		Extensive program design and implementation resources?						
		Deep research and development resources?						
		Structural response to innovation in place?						
		Multiple sources of innovation?						
		Innovation encouraged?						
		Organization is used to/embraces changes?						
		Other queries, identified by organization						

FIGURE B.3. ANALYSIS OF THE INTERNAL ORGANIZATIONAL CLIMATE. (continued)

Major Section		Important Criteria		Criteria Scores Score (0–4, 4 = high)				Comments
Number	Name	Description		Your Organization				
	a		b	c	d	e	f	

Customer focus:
  Quick response to customer complaints?
  Convenient manner for customers to register comments?
  Measure customer satisfaction?
  Evaluate competitor customer data?
  Extensive service training for front-line staff?
  "Secret shopper" program exists?
  Leadership has experienced services as customers?
  Other queries, identified by organization

Outcome measurement:
  Collect and analyze internal data?
  Collect and analyze external data?
  Use comparative databases?
  Benchmark, internally and externally?
  Extensive goal setting for leadership and staff?
  Use standardized data formats?
  Use evaluation and performance measurement systems?
  Compensations/rewards connected to organizational goals?

FIGURE B.3. ANALYSIS OF THE INTERNAL ORGANIZATIONAL CLIMATE. (continued)

Major Section		Important Criteria			Criteria Scores Score (0–4, 4 = high) Your Organization			Comments
Number	Name	Description	b	c	d	e	f	
a								

Description			
Other queries, identified by organization			
Creativity valued:			
Organization has a tolerance for misfits?			
Seek information from outside organization/industry?			
New projects encouraged?			
New projects nurtured/allowed to mature?			
Other queries, identified by organization			
Inquisitive learning:			
Continuing education is supported?			
External workshops/conferences are used?			
Learning resources are assembled and accessible?			
Other queries, identified by organization			
Acceptable risk:			
Failure is tolerated, even celebrated?			
Low fear of new project assignments?			

FIGURE B.3.   ANALYSIS OF THE INTERNAL ORGANIZATIONAL CLIMATE. (continued)

Major Section		Important Criteria	Criteria Scores Score (0-4, 4 = high)						Comments
Number	Name	Description	Your Organization						
a		b	c	d	e	f			
		Investment in R&D is visible?							
		Risk-taking is rewarded (career advancement)?							
		Risk-taking celebrated (heroes and myths)?							
		Other queries, identified by organization							
		Stretch goals:							
		Incremental progress, alone, is not tolerated?							
		Breakthrough strategy is embraced?							
		Bold, new initiatives are successfully launched?							
		Ambitious goal-setting throughout organization?							
		Leaders and staff accountable for goal attainment?							
		Other queries, identified by organization							
		Other priority criteria, identified by organization							

FIGURE B.3. ANALYSIS OF THE INTERNAL ORGANIZATIONAL CLIMATE. (continued)

Major Section		Important Criteria			Criteria Scores Score (0–4, 4 = high) Your Organization			Comments
Number	Name	Description						
	a		b	c	d	e	f	

Measurement

Direction-focused:
- Progress toward vision is measured in outcomes?
- Outcomes are defined by customer requirements?
- Financial information does not drive strategy?
- Information systems serve core processes?
- Other queries, identified by organization

Strategy inspired:
- Strategic plans guide daily activities?
- Outcome measures reflect strategic goals?
- Priorities are established and measured?
- Measures that no longer apply to strategic goals are eliminated?
- Other queries, identified by organization

Values/ethics:
- Outcomes reflect organizational values?

FIGURE B.3. ANALYSIS OF THE INTERNAL ORGANIZATIONAL CLIMATE. (continued)

Major Section		Important Criteria	Criteria Scores Score (0–4, 4 = high)			Comments
Number	Name	Description	Your Organization			
	a		b	c		
				d	e f	
		Financial results not equal to high value?				
		Profitability is an outcome, not a goal?				
		Other queries, identified by organization				
		Financial/time data:				
		Provide information for decision making?				
		Data do not drive decision making?				
		Data are shared freely internally, not hoarded?				
		New programs/processes supported to maturity?				
		Realistic development time is allocated?				
		R&D supported by resource allocation/recognition?				
		Reinforcements/rewards not tied solely to financial results?				
		Other queries, identified by organization				
		Other priority criteria, identified by organization				

FIGURE B.3. ANALYSIS OF THE INTERNAL ORGANIZATIONAL CLIMATE. (continued)

Major Section		Important Criteria	Criteria Scores Score (0–4, 4 = high)				Comments
Number	Name	Description	Your Organization				
	a		b	c	d	e	f	

Leadership

Behavioral base:
Leaders at all levels have expected behaviors defined?
Deviations from expected behaviors attended to with alacrity?
Leadership behaviors reinforce organizational alignment?
Other queries, identified by organization

Vision and direction:
Leaders create and communicate vision?
Leaders provide consistent direction for vision fulfillment?
Leaders model values supporting organizational direction?
Vision-based leadership inspires stakeholder commitment?
Other queries, identified by organization

Energy and direction:
Leaders actively delegate and hold subordinates accountable?
Leaders bring energy and enthusiasm to followers?

FIGURE B.3. ANALYSIS OF THE INTERNAL ORGANIZATIONAL CLIMATE. (continued)

Major Section		Important Criteria			Criteria Scores Score (0–4, 4 = high) Your Organization			Comments
Number	Name	Description			Score			
	a		b	c	d	e	f	
		New initiatives/projects supported with leadership energy?						
		Other queries, identified by organization						
		Key process focus:						
		Leaders remember/stay in touch with core business processes?						
		Focus on detail when critical processes are engaged?						
		Able to easily re-engage at critical junctures in process?						
		Other queries, identified by organization						
		Quality of worklife:						
		Leaders create an acceptable work environment?						
		Standards/values guide employment practices/procedures?						
		Strict implementation of employment policies/procedures?						
		Genuine celebration of achievement?						
		Rewards/recognition is for success at all levels?						
		Open and continuous communication is fostered?						
		Solicit and use employee satisfaction (climate) data?						

FIGURE B.3. ANALYSIS OF THE INTERNAL ORGANIZATIONAL CLIMATE. (continued)

Major Section		Important Criteria	Criteria Scores Score (0–4, 4 = high)					Comments
Number	Name	Description	Your Organization					
a			b	c	d	e	f	

Other queries, identified by organization

Other priority criteria, identified by organization

Other queries, identified by organization

Paradoxical

Balanced paradox:
Able to maintain broad focus and detailed implementation?
Financial outcomes balanced with priority for vision attainment?
Other queries, identified by organization

Conflict management:
Conflict in ideas is sought?
Personal attacks are not tolerated?
Respect for controversy is maintained?
Differing opinions and evaluation are encouraged?
Appropriate conflict resolution behavior is engaged?
Other queries, identified by organization

FIGURE B.3. ANALYSIS OF THE INTERNAL ORGANIZATIONAL CLIMATE. (continued)

Major Section		Important Criteria	Criteria Scores Score (0–4, 4 = high)				Comments
Number	Name	Description	Your Organization				
	a		b	c	d	e	f	

Dignity/respect:
Alternatives not selected based on political stature?
Favor-seeking discouraged?
Advancement is on merit, not political savvy?
Other queries, identified by organization

Other priority criteria, identified by organization

Deep Culture
Intensive values:
Values are palpable?
Behavior matches spoken values?
Ethical breeches quickly and appropriately handled?
Other queries, identified by organization

Walking the talk:
All leaders conform to values/ethics statement?
Goals, strategies, objectives reflect alignments with values?

FIGURE B.3. ANALYSIS OF THE INTERNAL ORGANIZATIONAL CLIMATE. (continued)

Major Section		Important Criteria		Criteria Scores Score (0–4, 4 = high)			Comments
Number	Name	Description		Your Organization			
	a		b	c	d	e	f

Other queries, identified by organization

Stakeholder valued:
All organizational members treated with respect and concern?
Concern for impact on stakeholders part of decision making?
Low or no necessity to seek collective bargaining protections?
Terminology used in referring to stakeholders reflects values?
Other queries, identified by organization

Evidence culture alignment:
All products reflect consistent adherence to culture?
Consistency between mission/ vision and strategy?
Publications/public pronouncements reflect values?
Policies/procedures reflect values?
Visible symbols of organization consistent with culture?
New members imbued with values/ethics in orientation?
Behaviors match expectations set for all?

FIGURE B.3. ANALYSIS OF THE INTERNAL ORGANIZATIONAL CLIMATE. (continued)

Major Section		Important Criteria	Criteria Scores Score (0–4, 4 = high) Your Organization			Comments
Number	Name	Description	d	e	f	
a		b c				

Measurement of outcomes is appropriate to goals/objectives?
Rewards match behaviors sought?
Other queries, identified by organization

Other priority criteria, identified by organization

Stretch Goals
Broad vision and ambition:
Progress is by breakthrough?
Radical growth, or unique strategy sought?
Rewards in place for exceeding expectations?
Energy/competitive drive focused on market?
Other queries, identified by organization

Thriving:
Goal is not survival, but to thrive?
Financial outcome measures reinforce expectation of success?
Small improvement supported, superseded by breakthroughs?

FIGURE B.3. ANALYSIS OF THE INTERNAL ORGANIZATIONAL CLIMATE. (continued)

Major Section		Important Criteria			Criteria Scores Score (0-4, 4 = high)			Comments
Number	Name	Description			Your Organization			
	a		b	c	d	e	f	
		Little tolerance for mediocrity? Other queries, identified by organization						
		Externally competitive: Internal competitive behavior is discouraged? Competitive focus is on market rivals? Benchmarking/best in field practices supported? Measurement against top performers expected? Other queries, identified by organization						
		Leadership/information and analysis: Work processes are supported by information technology? Information is exchanged quickly and easily? Information is available where and when needed? Other queries, identified by organization						
		Other priority criteria, identified by organization						

FIGURE B.3. ANALYSIS OF THE INTERNAL ORGANIZATIONAL CLIMATE. (continued)

Major Section		Important Criteria	Criteria Scores Score (0–4, 4 = high)				Comments
Number	Name	Description	Your Organization				
	a	b	c	d	e	f	

Execution With Excellence

Quality obsessed:
- Mediocre performance or results not tolerated?
- Goals for quality continuously increased?
- Measurement of quality emphasized?
- Star performers/teams celebrated?
- Poor performance corrected and root causes addressed?
- Other queries, identified by organization

Measurement focused:
- Performance parameters are set, known, assessed for all?
- Continuous quality improvement practiced throughout organization?
- Departmental goals assessed?
- Organizational goals assessed?
- Other queries, identified by organization

Breakthrough strategy:
- Continuous improvement of core processes expected?

FIGURE B.3. ANALYSIS OF THE INTERNAL ORGANIZATIONAL CLIMATE. (continued)

Major Section		Important Criteria	Criteria Scores Score (0–4, 4 = high)				Comments
Number	Name	Description	Your Organization				
	a		b	c	d	e	f	
		Radical plus incremental improvement is sought?						
		New programs/projects supported and implemented?						
		R&D, hunting parties, innovation encouraged?						
		Other queries, identified by organization						
		Obsolescence aware: Organization does not rest on past achievements?						
		Current innovation is more important than past?						
		Don't believe in the myth of invulnerability?						
		"What have we done for customers lately" is continuous question?						
		Other queries, identified by organization						
		Other priority criteria, identified by organization						

REFERENCES

Alvarez-Buylla, Yvette, Hamilton, Jennifer, and Korkowski, Melanie. *Analysis of Access to Health Care.* IOE 481 project report. Ann Arbor: Industrial and Operations Engineering, University of Michigan, Dec. 10, 1996.

American Hospital Association. *Hospital Statistics 95/96.* Chicago: American Hospital Association, 1995.

Auerbach, Stuart. "Falling Health Care Employment Could Hurt Economy, Experts Warn." *Washington Post,* July 24, 1996.

Barker, Joel Arthur. *Discovering the Future: The Business of Paradigms.* St. Paul, Minn.: ILI Press, 1989.

Barker, Joel Arthur. *Future Edge.* New York: Morrow, 1992.

Beckham, Daniel J. "The Vision Thing." *Healthcare Forum Journal,* Mar.–Apr. 1994, pp. 60–68.

Berman, Steve (ed.). "Using Malcolm Baldrige National Quality Award Criteria for Improvement: An Interview with Ellen Gaucher." *Journal on Quality Improvement,* 1995, *21*(5), 249–256.

Chappell, Tom. *The Soul of a Business.* New York: Bantam, 1993.

Chawla, Sarita, and Renesch, John (eds.). *Learning Organizations: Developing Cultures for Tomorrow's Workplace.* Portland, Ore.: Productivity Press, 1995.

Codrescu, Andrei. *The Dog with the Chip in His Neck.* New York: St. Martin's Press, 1996.

Coffey, Richard J., Othman, J. Elizabeth, and Walters, Janet I. "Extending the Application of Critical Path Methods." *Quality Management in Health Care,* 1995, *3*(2), 14–29.

Coffey, Richard J., Richards, Janet S., Wintermeyer-Pingel, Susan A., and Le Roy, Sarah S. "Critical Paths: Linking Outcomes for Patients, Clinicians, and Payers." In Peter R. Kongstvedt (ed.), *The Managed Health Care Handbook.* (3rd ed.) Gaithersburg, Md.: Aspen, 1996.

Coffey, Richard J., and others. "An Introduction to Critical Paths." *Quality Management in Health Care,* 1992, *1*(1), 45–54.

Collins, James C., and Porras, Jerry I. *Built to Last.* New York: HarperBusiness, 1994.

Commission on Accreditation of Rehabilitation Facilities. *1996 Standards Manual and Interpretive Guidelines for Behavioral Health.* Tucson, Ariz.: Commission on Accreditation of Rehabilitation Facilities, 1996.

Commission on Professional and Hospital Activities. *International Classification of Diseases, Ninth Revision, Clinical Modification.* Ann Arbor, Mich.: Commission on Professional and Hospital Activities, 1979.

Council on Accreditation of Services for Families and Children. *Standards for Agency Management and Service Delivery.* New York: Council on Accreditation of Services for Families and Children, 1996.

Fenner, Peter, and Fenner, Kate. *Manual of Nurse Recruitment and Retention.* (2nd ed.) Gaithersburg, Md.: Aspen, 1989.

Gaucher, Ellen J., and Coffey, Richard J. *Total Quality in Healthcare: From Theory to Practice.* San Francisco: Jossey-Bass, 1993.

Gaucher, Ellen J., and Coffey, Richard J. *From Incremental Improvement to Breakthrough Performance.* Unpublished manuscript, 1997.

Gonen, Julianna S., and Probyn, Susan L. "The Evolution of Accreditation." *HMO Magazine,* Jan.–Feb. 1996, pp. 53–57.

Gouillart, Francis J., and Kelly, Charles M. *Transforming the Organization.* New York: McGraw-Hill, 1995.

Griffith, John R., Warden, Gail, Dowling, William L., and Pelham, Judith C. "Managing the Transition to Integrated Health Care Organizations." *Frontiers of Health Services Management,* 1996, *12*(4), 4.

Health and Welfare Canada. *A Taxonomy of the Canadian Health Care System.* Developed by Richard J. Coffey. Ottawa: Health and Welfare Canada, 1979.

Healthcare Forum. *Risky Business: Mastering the New Business of Health.* Strategic Simulation Series. San Francisco: Healthcare Forum, 1996.

Hesselbein, Frances, Goldsmith, Marshall, and Beckhard, Richard (eds.). *The Organization of the Future.* San Francisco: Jossey-Bass, 1997.

Jackson, Francis W. *The Malcolm Baldrige National Quality Award Criteria/Joint Commission Accreditation Standard Crosswalk.* Kingsport, Tenn.: Bishop Associates, 1995.

Japsen, Bruce. "Not-for-Profit Deal Draws Fire." *Modern Healthcare,* Dec. 9, 1996, pp. 4–5.

Joint Commission on Accreditation of Healthcare Organizations. *Lexikon: Dictionary of Health Care Terms, Organizations, and Acronyms for the Era of Reform.* Oakbrook Terrace, Ill.: Joint Commission on Accreditation of Healthcare Organizations, 1994.

Joint Commission on Accreditation of Healthcare Organizations. *Comprehensive Accreditation Manual for Hospitals: The Official Handbook.* Oakbrook Terrace, Ill.: Joint Commission on Accreditation of Healthcare Organizations, 1996.

Kano, Noriaki. "Attractive Quality and Must-Be Quality." Paper presented at the Twelfth Annual Meeting of the Japan Society of Quality Control, 1982. Translated by GOAL/QPC. Methuen, Mass.: GOAL/QPC, 1984.

Kaplan, Robert S., and Norton, David P. "Using the Balanced Scorecard as a Strategic Management System." *Harvard Business Review,* Jan.–Feb. 1996, pp. 75–85.

Kotter, John P., and Heskett, James L. *Corporate Culture and Performance.* New York: Free Press, 1992.

Lamm, Richard D. "The Ten Commandments of Health Care." In P. R. Lee and L. Estes (eds.), *The Nation's Health*. (3d ed.) Boston: Jones & Bartlett, 1990.

Malcolm Baldrige National Quality Award Office at the National Institute of Standards and Technology. *Health Care Pilot Criteria 1995*. Gaithersburg, Md.: National Institute of Standards and Technology, 1995.

Marszalek-Gaucher, Ellen, and Coffey, Richard J. *Transforming Healthcare Organizations: How to Achieve and Sustain Organizational Excellence*. San Francisco: Jossey-Bass, 1990.

National Committee for Quality Assurance. *NCQA Accreditation Manual*. Washington, D.C.: National Committee for Quality Assurance, 1994.

National Committee for Quality Health Care and the Lewin Group. *Tracking the System*. Washington, D.C.: National Committee for Quality Health Care, 1997.

Peters, Thomas J., and Waterman, Robert H. *In Search of Excellence: Lessons from America's Best-Run Companies*. New York: HarperCollins, 1982.

Ryan, Kathleen D., and Oestreich, Daniel K. *Driving Fear Out of the Workplace: How to Overcome the Invisible Barriers to Quality, Productivity, and Innovation*. San Francisco: Jossey-Bass, 1991.

Shortell, Stephen M., and others. *Remaking Health Care in America: Building Organized Delivery Systems*. San Francisco: Jossey-Bass, 1996.

Spector, Robert, and McCarthy, Patrick D. *Nordstrom Way*. New York: Wiley, 1995.

Stefl, Mary E. "Editorial Page." *Frontiers of Health Services Management*, 1996, *12*(4), 1.

U.S. Department of Commerce, Bureau of the Census. "Health Insurance Coverage, 1995." *Current Population Reports*. Household Economic Studies, P60–195. Washington, D.C.: U.S. Government Printing Office, 1996.

Uslaner, Eric. *The Decline of Comity in Congress*. Ann Arbor, Mich.: University Press, 1993.

Young, Gary J., and Coffey, Richard J. "Aligning Employees Around Quality Improvement Goals: The Role of Performance Agreements." In *Proceedings of the 1997 NSF Design and Manufacturing Grantees Conference*. Arlington, Va.: National Science Foundation, 1997.

INDEX

A

Access: expectations on, 109; issues of, 21–24, 198
Accreditation Association for Ambulatory Health Care, 98
Actions: and external expectations, 136–137; and internal organizational climate, 164–165; and scope of services, 89–90
Advocacy focus, 57
Agency for Health Care Policy and Research, 101
Alliance for Alternatives in Health Care, 101
Alliances, 4
Allied health workers, 65–66
Alvarez-Buylla, Y., 24n
Ambulatory settings, 62
American Association of Retired Persons, 100
American Hospital Association, 2
American Managed Care and Review Association, 100
Analyses: aspects of, 27–35; and balanced scorecard, 183–184; combined, 184–192; compar-

isons for, 31–32; components of, 27–28; criteria for, 185; of external expectations, 28–29, 91–137; final, 183–192; of internal organizational climate, 28–29, 138–165; level of detail for, 32–33, 184–185; scope of, 30–33; of scope of services, 28, 73–90; scoring, 186–187, 190, 191; staffing for, 33–34, 74–75, 125, 156; types of, 28–30; uses of, 34–35; weights in, 186, 187, 190, 191
Analysis tool: for external expectations, 116–123, 125–127; for internal organizational climate, 156–160; for scope of services, 77, 79–88; scoring system for, 79, 126–127, 160; steps for using, 88–89, 134, 163–164
Appropriateness: analyses of, 27–28, 30, 35; of external expectations, 124, 133–134, 136–137; of internal organizational climate, 155–156, 165; of scope of services, 74, 90; and taxonomy, 69, 71

Areawide settings, 60
Arizona: safety in, 52; social conditions in, 40
Assistive caregivers, 66
Auerbach, S., 3

B

Balance issues: for analyses, 183–184; for health and social initiatives, 21; of paradox and conflict, 147–148
Berman, S., 95
Biomedical and consumer product safety focus, 51–52
Blue Cross, 92
Board composition, 168, 172–174, 182
Buildings, as resource, 67

C

California, social conditions in, 40
Canada, access and scope in, 21
Care continuum expectations, 109–110, 119–121, 131
Career Architect, 171